During the Great Depression, a small group of farmers from the Texas Panhandle migrated to western New Mexico and set up a new community. They were homesteaders, the last of a long procession that settled the frontier. In the remote village which they founded, many of the basic American frontier values are still perpetuated: hopeful mastery over nature; optimistic faith in an ever brighter future; "rugged individualism" as a pattern preferred to cooperative social effort; and a complex system of group feelings of inferiority and superiority.

The author of this book, and several fellow research workers, lived in "Homestead" over a period of five years, participated in community affairs, gained the confidence of the "Homestead" people. They studied the values of the "Homesteaders" as well as the dramatic changes taking place in the relationship between this small, homogeneous village and the larger Southwestern culture which surrounds it.

Mr. Vogt shows how the same values which made possible the settlement and early growth of "Homestead" are now accelerating a change to a kind of community which the "Homesteaders" neither anticipated nor desired — in fact believed tural change, this book will be of lasting interest to anyone interested in American culture and values as well as to professional social scientists. Evon Z. Vogt, co-author of **Navaho Means People** (Harvard University Press), is Associate Professor of Social Anthropology at Harvard University.

MODERN HOMESTEADERS

MODERN HOMESTEADERS

**THE LIFE OF A TWENTIETH-CENTURY
FRONTIER COMMUNITY**

EVON Z. Zartman VOGT

THE BELKNAP PRESS
of HARVARD UNIVERSITY PRESS · CAMBRIDGE · 1955

© Copyright, 1955, by the President and Fellows of Harvard College

Distributed in Great Britain by
Geoffrey Cumberlege, Oxford University Press, London

Printed in the United States of America
Library of Congress Catalog Card Number 55–10978

TO
NAN

Preface

In the summer of 1949 the Laboratory of Social Relations of Harvard University initiated a six-year field study of the value systems of five distinct cultural groups in western New Mexico: Navaho, Pueblo,* Spanish-American, Mormon, and Texan homesteader. This project, known as "The Comparative Study of Values in Five Cultures," was financed by a grant from the Social Science Division of the Rockefeller Foundation. The basic design of the project was based upon the fact that the five distinct cultural groups coexist in the same relatively small ecological area in New Mexico, yet have developed over time and continue to maintain distinct value systems. In this field laboratory, project research workers from several social science disciplines have worked on the development and refinement of concepts, methods, and research tools which we hope will be applicable to other areas of the world where diverse cultures are in intimate contact. They have also added significantly to ethnographic knowledge of the Southwest because the research area contains a cross section of Southwestern cultural types (Pueblo and Athabascan Indian, Spanish-American, and two variants of generalized American culture: the Mormons and the Texan homesteaders).†

This book on the Homesteaders is the first major publication on the value system of the Texan homesteader culture in our project research area. Although the study singles out the Homesteaders for special attention, I have made comparisons with the other cultures in the area when this was important to my analysis, and I have at-

---

* "Pueblo" is a fictitious name for a Southwestern Indian community having a pueblo-type culture.

† For a more complete description of the project, see the Preface by Clyde Kluckhohn to Evon Z. Vogt, *Navaho Veterans: A Study of Changing Values*, Peabody Museum of Harvard University, Papers, vol. 41, no. 1 (1951); for a list of the project's publications, see the front matter in Watson Smith and John M. Roberts, *Zuni Law: a Preliminary Study in Values*, Peabody Museum of Harvard University, Papers, vol. 43, no. 1 (1954).

tempted to maintain as much of the inter-cultural context as possible. With the eventual publication of additional books, monographs, and articles on the Homesteaders, on the other cultures of the area, and the publication of our final over-all report on the project research, the interested reader will be provided with a deeper and far richer picture of our project and of the interesting research area than it is possible to present in this small volume.

My field work in the Homestead community was initiated in October 1949 when I moved into the village with my wife and three children. We lived in Homestead until September 1950, and we returned again in the summers of 1951 and 1952, making a total of eighteen months of actual residence and active field research in the community. During the period from 1949 to 1954, there were fifteen other field workers engaged in research in Homestead on various problems and for various periods of time. But unless otherwise specified, statements about the "current" or "present" situation in Homestead refer to 1949–1950, the period of my full year of field work in the community.

Since I had previously lived in the vicinity of Homestead, our first contacts with the people were ones of renewing old acquaintanceships and making new ones. We lived in a rented farm house located one-half mile from the village center. This location proved to be ideal since we were close enough to the center of activities to make daily observations, but far enough away to have days to ourselves to spend in writing, reading, and study.

From the first, the Homesteaders were interested in our research, which was explained to them frankly in terms of making a complete study of the history, geography, and social life of the community. And as some people in Homestead were in the farming business, others in the store business, we were defined as being in the "research business." Families from all parts of the community were kind enough to include us in their social activities. It was apparent that the community as a whole welcomed research workers and expressed both feelings of pride that their village had been selected for investigation and feelings that the presence of field workers "livened up the community."

During the first four months in Homestead we restricted our activities to participant observation of the uniformities of responses which occurred in diverse situations within the community. We

PREFACE

made an effort to visit with every family in the community, to attend both churches, all dances, and every meeting that took place for whatever purpose. Inter-dining relationships were established with ten different families in various parts of the community. We also made on-the-spot observations and did later interviewing on all crisis and conflict situations which occurred.

In January 1950, I selected a strategic sample of twenty informants and began a formal program of intensive interviewing which continued into the summer of 1950, along with participant observation activities. Informants were not paid for these interviews. In Homestead, "talking" is defined as loafing rather than working behavior, and to be paid for an interview would be considered most inappropriate. Instead, we attempted to even up the score somewhat by giving gifts, by doing favors for the Homesteaders when we went to Gallup or other cities, and by inviting them to meals in our home.

In the spring of 1950, I also collected autobiographies from each of the sixteen high-school students (written during their English classes), for which I paid them ten cents a page. In addition, the records of the school, the Farm Bureau, and the Homestead Water Coöperative were given to me to examine, and relevant data were recorded in the field notes. Copies of the high-school paper published during the 1930's provided particularly rich historical data. Interviews with government officials in the local offices of the Departments of Agriculture and the Interior in Albuquerque also provided highly useful information on Homestead from the point of view of the government programs that have been attempted to benefit the people of the community.

Finally, during 1951–1952 I became, at the request of the community, an advisory member of the Homestead Road Committee, whose function was to promote the construction of a paved highway through the community. The economic data which I had collected on Homestead was utilized by the committee to support the community's request to the county and state highway officials. Participation on this committee also gave me the opportunity to observe the responses of the Homesteaders in their meetings with other peoples in the area and with the highway officials.

I am deeply aware of the fact that the study of Homestead was a collaborative effort involving the time, energy, and intellectual resources not only of the members of the project research program but also of the residents of Homestead. My study has benefited greatly from the use of unpublished field materials gathered by other project workers, including Barbara Chartier Ayres, Wilfrid C. Bailey, Mary Louise Bensley, Eleanor Hollenberg Chasdi, Munro S. Edmonson, Helen Faigin, Florence Kluckhohn, Edgar L. Lowell, Donald N. Michael, Paul Sears, Margaret Sperry, Fred L. Strodtbeck, Irving Telling, Wayne W. Untereiner, Otto von Mering, and Morris Zelditch, Jr.

The manuscript, or parts of it, in various stages of preparation had the benefit of a critical reading and helpful suggestions from Ethel M. Albert, J. O. Brew, Wilfrid C. Bailey, Clyde Kluckhohn, Florence Kluckhohn, Robert C. Mendenhall, Donald N. Michael, Thomas F. O'Dea, Anne Parsons, Talcott Parsons, John M. Roberts, David Schneider, Bernard J. Siegel, Fred L. Strodtbeck, Wayne W. Untereiner, and Otto von Mering. The criticisms of Ethel M. Albert, J. O. Brew, Clyde Kluckhohn, Talcott Parsons, and David Schneider were particularly detailed, and I am most grateful for the time they devoted to the task. Final responsibility for what I have written of course rests with me, for I was often stubborn in not following what may turn out to have been good advice.

Throughout the period of the research and during the preparation of this manuscript I have received the fullest coöperation and encouragement from the members of the Project Advisory Committee, Professors J. O. Brew, Clyde Kluckhohn, and Talcott Parsons; the Director, Professor Samuel A. Stouffer, and the Associate Director, Professor Richard Solomon, of the Laboratory of Social Relations; and Professor John M. Roberts, who was the Coördinator of the project from 1949 to 1953.

I am also indebted to Clarissa Schnebli for careful and painstaking editorial criticism of the manuscript, to June Barnes for the hours she devoted to typing on the manuscript in the project office, and to James Gazaway for bibliographic work.

The photographs of Homestead were taken by David DeHarport in the summer of 1953, and I wish to express my appreciation to him for permission to use them in the book.

I owe a great debt to my wife, Naneen Hiller Vogt, to whom the book is dedicated, for her consummate skill and patience in being a wife, a mother, and helping me with the field work all at the same time.

PREFACE

Finally, I wish to express my appreciation to our Homesteader friends who gave us so much of their time, interest, and coöperation. Many of the Homesteaders are not going to like the conclusions I have drawn in this book about their community, but I am certain they will understand that I nevertheless have the greatest respect and admiration for their courage and resourcefulness and that I regard them as being among my closest personal friends.

Evon Z. Vogt

Cambridge, Massachusetts
June 1955

CONTENTS

Introduction	1

PART I: THE SITUATION

1. **The Geography and Demography of Homestead**	23
2. **The Economy of Homestead**	37

PART II: THE VALUES AND THE COMMUNITY

3. **Hopeful Mastery Over Nature**	63
4. **Living in the Future**	93
5. **Working and Loafing**	109
6. **Group Superiority and Inferiority**	122
7. **The Atomistic Social Order**	140
8. **Conclusions**	173
Appendix	191
Notes	211
Index	229

TABLES

Table I. Annual precipitation and bean production in Homestead — 29

Table II. Emigration from Homestead by family — 101

Table III. Occupations of young people who have left Homestead — 103

GRAPHS

Immigration to and emigration from Homestead by families — 33

Age and sex distribution in Homestead — 35

MAP

Sketch map of the Homestead area — 19

ILLUSTRATIONS

Twentieth-Century Homesteader Farms Beans with Modern Tractor in an Arid Land. — 44

Incomplete Gymnasium Stands as a Monument to Rugged Individualism. — 45

The Village Store. — 45

The Homestead Bar Is a Popular Loafing Place — by Custom for Men Only. — 76

As Population Declines, Former Businesses Close Up. — 76

Village Store Is a Place of Business — and Visiting. — 77

Summer Rodeos are Favorite Pastimes. — 108

Young Homesteaders Begin to Work in the Bean Fields at an Early Age. — 109

No Stained Glass Windows — but Services Take Place Here Every Sunday. — 140

Modern New Mexico Is the Land of the Pickup Truck. — 140

Homesteader Houses in the Pinyon Trees. — 141

MODERN HOMESTEADERS

Introduction

This book is a study of a small, dry-land, bean-farming community in western New Mexico — a community which is located far from centers of economic and political power. What happens to it is certainly unimportant in the course of world events, however important it may seem to the 200 people who live in this village which I shall call Homestead.* However, the study of Homestead is of special interest because of the extent to which certain core American values are manifested in the lives of these twentieth-century homesteaders and because of the critical role which these values have played in the settlement, development, and subsequent transformation of the community. I shall attempt to describe and document these values and then demonstrate that while a cluster of certain American frontier values had a positive effect upon the settlement and early development of Homestead, the continuing adherence to these same values in the face of a changing environmental and economic situation is now contributing to the transformation of the community from a farming village into a region of widely scattered small ranches.

It will be noted from the table of contents that the organization of this book represents a marked departure from the traditional organization of an anthropological study with the familiar categories of "economics," "social structure," "religion," etc. Because I am convinced that the course of events in Homestead can best be understood in terms of values, I am presenting the description of this particular community within the framework of value-orientations. As such, the book will constitute an experiment in method of presentation and in theoretical treatment.

There are a number of alternative hypotheses which one might attempt to apply as a basic explanation of the cultural patterns and

* All specific place names, including "Homestead," in the immediate research area are fictitious in order to protect the anonymity of my informants.

processes described in this study: (a) that these cultural features are determined primarily by the environmental situation; (b) that they are determined primarily by considerations of biological and/or social survival; (c) that these features are primarily a function of basic economic realities; or (d) that they are determined by some combination of these various factors.

The comparative data of the Values Study and the body of data collected on Homestead offer convincing evidence for the superior adequacy of the central hypothesis of this study: that the value-orientations are not merely dependent variables but have been one of the significant determinants in the course of events throughout the history of the community.

If the patterns and processes were determined by situations, primarily or exclusively, then all of the five neighboring cultural groups (Navaho, Pueblo, Spanish-American, Mormon, and Texan homesteader) under scrutiny by our project in western New Mexico would manifest markedly similar cultures arising from the relatively uniform environmental situation. Since this is not the case, hypothesis *a* cannot be correct.

If the cultural patterns and processes were determined primarily by considerations of sheer biological and/or social survival, i.e., if these patterns and processes were tension-reducing for individuals and had functional utility for the preservation of the society, then the Homesteaders would abandon their present dry-land bean-farming adjustment to the situation and turn to a different crop or leave for better land or different jobs.[1] Since the majority follow none of these alternatives (see Chapters 2, 5, and 7), hypothesis *b* can hardly be correct.

If the patterns and processes in the cultural system were determined by rational economic factors, the Homesteaders would likewise turn from bean farming to other economic pursuits (see Chapter 2). Again, since most of them do not shift to more secure economic alternatives, the hypothesis of economic determinism cannot be correct.

Even if one considers all these sets of factors in various combinations, there is a large residue of unexplained behavioral facts. These can be meaningfully treated in terms of the central value-orientations of the community.

INTRODUCTION

DEFINITIONS AND CONCEPTS

What are "value-orientations" and how does one describe them?[2]

In arriving at a definition of "value-orientation," one may begin by imagining what human life might be like without culture. This kind of speculative inquiry can be made only on the basis of scattered bits of evidence, such as observations of the behavior of infants at birth, or of the behavior of individuals in a situation where the social and cultural controls have disintegrated. This evidence strongly suggests that cultureless human action tends to be trial-and-error or "instinct-driven." Once we add culture, however, the random characteristics give way to patterned regularities. This can be observed on the one hand in the socializing of infants as they gradually acquire culturally regulated ways of behaving. The reverse process can be observed in situations of cultural disorganization where the cultural controls break down and the behavior tends to be normless and more random again.

One may then ask why it is that men living in groups with a culture manifest behavior which is patterned rather than random. Two fundamental observations, based upon our current theoretical knowledge and empirical evidence, may be made in answer to this question. Although each culture *is in all its concreteness* unique (like a single snowflake), the earth's cultures do exhibit basic uniformities. The second observation is that overlying these uniformities there are important variations from culture to culture. It is as if the potential randomness of behavior were patterned on the one hand by certain *universal* properties of human groups (like the minimum mechanical requirements for any clock which runs and keeps reasonably good time), and on the other by *particular* properties of specific cultural groups (like the technical elaborations that characterize particular types of clocks).

Our accumulating evidence suggests that the universal properties of human groups stem from the biological and social nature of man and from the general characteristics of the earth's natural environment. There is no need here to outline the familiar facts about human biology, but I might note by way of illustration three fundamentals which underlie cultural uniformities. First, the dependence upon the mammalian mother for physical nourishment and for emotional support until the human young become adults capable of self-support makes necessary a process of socialization which exhibits certain similarities regardless of cultural type.[3] Second, the fact that there are two sexes with different functions in the reproductive cycle has certain universal consequences for human cultures.[4] Third, the fact of eventual death for all humans requires an adjustment by the aging members of our species and a ritual readjustment of the survivors when death occurs, which again has certain basic universal regularities.[5]

The social nature of man also necessitates certain fundamental prerequisites for the continuing existence of any human group. These have been most clearly traced out by Talcott Parsons,[6] who indicates that the following basic problems must be met by any society: provision for the minimum biological and psychological needs of a sufficient proportion of the society's members; coördination of the activities of the various individual members so that there will be a minimum degree of order in the society; adequate motivation of a sufficient proportion of the members of the society to perform the essential social roles.

The ecological situations of man affect both the universal and the particular properties of the world's cultures. Certain obvious uniformities in human behavior clearly stem from such fundamental factors as gravity, the presence of both plant and animal life in most parts of the world, precipitation in the form of rain or snow. On the other hand, it is equally clear that ecological situations vary significantly over the world, setting up different types of limits on the kinds of cultures possible in any given region.

But within the limits imposed by biology, by society, and by a given type of ecology, there are alternative lines of action which are followed by specific groups possessing, as a result, distinctive and particular properties. Although there exists the possibility that some of the differences among cultures are due to genetic factors, a more crucial variable in accounting for the major differences would appear to be in the value-aspects of culture. We may say preliminarily that these value-aspects have to do with the *selecting, regulating,* and *goal-discriminating* processes in a cultural system.

The phenomenon which the anthropologist calls "culture" has been conceptualized in various ways,[7] but whatever view one holds, an important and distinctive aspect of culture is its *selectivity*, which often seems arbitrary. Certain "choices" that are not completely re-

INTRODUCTION

ducible to biological or social imperatives or to the physical environment seem to be made by human groups which "set off" the cultural process in one direction rather than another.[8] For example, the focus upon curing and the gradual elaboration and differentiation of ceremonials based upon curing in Navaho culture are phenomena that are not completely reducible to the situational requirements of Navaho life. Much of the specific content of these ceremonials ultimately derives from Pueblo ceremonialism, but the Navahos' re-structuring and elaboration of these rituals can be understood only in terms of the selecting value-orientations of the Navaho. It is as if the value-orientations of a cultural group function as selective "screens" or "filters" in the diffusion of cultural materials from other societies, and then as cultural "automatic pilots" which permit cultural changes and elaborations in certain directions but not in others.

The value-orientations which function as selectors in cultural processes are in part precipitates of historical events in the cultural history of any given group. They may have been influenced by particular types of historical contacts, by the idiosyncrasies of important individuals, by natural catastrophes the group has faced, or by combinations of all these. In a systematic analysis of the cultural history, it would be possible to trace the development of value-orientations in detail. Whatever their origins in the history of the group, however, the important point is that a set of distinctive value-orientations which grows out of past situations and is in turn fed back into the on-going cultural processes is carried by every cultural group.

Value-orientations are not only *selectors* which give a certain directionality to cultural processes. They also function as *regulators* in the system by continually defining and re-defining the limits of permissible behavior within any given role. In this function they provide the fundamental basis for a set of moral norms which support the crucial institutions of a culture. For example, the definition of the role of schoolteacher in Homestead always involves certain restrictions on behavior which is permitted other persons in the community. Teachers should not drink or play poker, women teachers are not permitted to smoke, and all are generally expected to be "good moral people" and moral exemplars for the young in the educational institutions of the community. Transgression of these moral dictates constitutes a flaunting of important values and inevitably results in action starting with gossip and culminating with a political movement to remove the teacher from his position.

Finally, value-orientations provide a set of criteria for the ranking or *discrimination of goals* for future action. That is, they are constantly reminding individuals or groups that certain goals are more important than others, or that goal X takes precedence over goal Y. For example, there is present in Homestead the familiar American goal conflict between "working hard to accumulate money and property" and "having fun." When the chips are down, however, the goal of accumulation takes precedence over the goal of recreation because of the value stress on personal achievement and "being a success." *

These *selective, regulative,* and *goal-discriminative* aspects of value-orientations are intimately interrelated, and can only be separated for analytical purposes. This is precisely why they are called "aspects" of value-orientations. Any given major value-orientation in a culture may be expressed by a single statement, but may be viewed analytically in all three of its functional aspects. For example, the Homesteaders described in this book are fundamentally committed to the notion that human relations are more desirable when organized on a basis of "rugged individualism" where "each man can be his own boss." This single statement of a major value-orientation of this cultural group can be seen as it is manifested in: (*a*) the *selection* of the pattern of individually owned farmsteads with each family living on its farm, rather than some system of coöperative land-management or, at the least, a settlement pattern in which the people live in a compact village and travel out to work their own farms each day (as the Mormons and Spanish-Americans do); (*b*) the *regulatory* functions provided by the definitions of appropriate behavior for an individual farmer in the community today which prescribe that continuing to live on isolated farms is more desirable than living "all bunched up the way them Mexicans do"; and (*c*) the *goal-discriminations* in the system which hold forth promise of

---

* It should be noted that statements to the effect that value-orientations "do" certain things or "function" in a particular way do not mean that I am reifying "values." These types of statements are shorthand ways of saying that individuals with certain motivations built into their personality systems by virtue of early socialization and later experience in a cultural group, behave in patterned ways which produce these selective, regulative, and other effects in the interactive processes in the group.

greater prestige to the young man (or to the tenant farmer) who works self-reliantly for his own gain, saves his money, and strives to purchase his own farm, than to the person who spends his money freely and is satisfied with wage labor or tenant farming as an occupational pursuit.

Finally, it should be emphasized that value-orientations are conceptualized as meaningful clusters of associated values and not just lists of things. This is not to say that there is of necessity complete consistency in the values of a culture. Indeed, it may well be, as Florence Kluckhohn suggests, that variant value-orientations are essential for the maintenance of a given cultural system.[9] It is clear, however, that value-orientations occur in patterned arrangements and that these arrangements may add as much to the unique qualities of a culture as the presence or absence of given values.

This clustering of values around certain foci also provides an important conceptual distinction between a "value" and a "value-orientation." "Values" are conceptions of the desirable which permeate an entire culture and may be treated as long lists ranging from the desirable ways of making pottery or arrowheads to the desirable ways of worshipping the gods; in fact, every cultural feature may have its value aspects. Value-orientations, on the other hand, are conceptualized as the patterned clusters of certain associated values around important foci in the life situation of a cultural group. They are, in Clyde Kluckhohn's words, "generalized and organized conceptions"[10] which provide customary orientations to such important foci in the human situation as man's relation to nature, to time, to work and play, to other cultural groups, and to his fellow men within the community.

DESCRIPTION OF VALUE-ORIENTATIONS

Since value-orientations are regarded as aspects of the patterning processes of culture, I began my study of Homestead with the premise that the methods that have long applied to the description of culture patterns also apply to the description of values. As far as I know, there is as yet no reliable short cut or "magical" way for discovering the value system of a culture. Although one may use questionnaires or projective tests of various types on samples of informants, or perform semicontrolled experiments to make one's statements about values more rigorous, it is important to take these steps only at points in the research enterprise where they are appropriate in the over-all plan of investigation. The data basic to the analysis must come from observing overt and verbal behavior either in recurrent situations in the social life of a people or induced in interviews.

In my judgment, the genuine innovations in methodology for values research in the current exploratory stages of inquiry are more likely to come by carefully singling out the types of behaviors and situations to be observed, the types of data to be gathered in the interview situation, the types of inferences which are made from these classes of behavior, and the kinds of predictions which are made about the course of events in a given culture. Once these basic steps are taken, it will be possible to bring any number of new and old techniques to bear upon the problem of actually collecting the data. If attention is centered prematurely upon research techniques, we are likely to be in the awkward position of putting a cart of techniques before the important methodological horse.

The type of situation which seems to offer the nearest approach to a focus for observation that is uniquely adapted to the study of values is the *choice-situation*. This consists essentially of a situation in which an individual, or group of individuals, is confronted with alternative possibilities and is asked (as in an interview or questionnaire) or is forced (by exigencies of a situation in real life where choice is necessary) to choose among the possibilities. Since, by definition, values influence choice-behavior, careful observation of the chosen response and of the reasons given therefor provides one fruitful approach to the problem of delineating the value system of a culture. This method can be utilized with a variety of specific techniques as, for example, the observation of groups responding to choice-situations, the focused interview, the questionnaire, the situational choice test, and the use of small-group discussions.

It should be emphasized that the choices themselves are not values; rather, the choices are made by the informants' balancing the situational requirements against their value-commitments. The values must be inferred from the choices made and from what the informants say about why they are making given choices.

The use of this method also raises the problem of how one knows that values of a certain kind are operative in a given situation and that the informants are not merely reacting to biological impulses,

INTRODUCTION

economic forces, or other situational requirements. There are at least three means by which an observer may control these situational variables. First, a careful study of the situation in which the responses occur may reveal that alternative possibilities are genuinely present. The choices observed may then be assumed to involve at least some value considerations. This presumption may be fortified by observation of the fact that some individuals or some groups within the cultural system, who do not vary significantly from the majority in biology or in social position, occasionally, or frequently, make other choices; this deviation will indicate that the situation in question is indeed a choice-situation. Second, controls can also be brought to bear upon the problem by comparing choice-behavior in two or more different cultural groups which live in comparable ecological settings, have comparable economic systems, and are confronted with identical problems: for example, the problem of drought. It may be assumed that the variations in response to the common problem result from differing sets of cultural values — and herein lies the methodological strength of the comparative design of the Values Study project. Third, the values postulated on the basis of observations made of certain past choice-situations, e.g., the choice of settlement pattern in Homestead, may be checked by predicting in advance what the group will do when a new comparable situation arises in the course of events, e.g., the problem of constructing a high-school gymnasium in Homestead, and then by making careful observations of what in fact does happen. In the example given, the settlement pattern in Homestead had led the observer to postulate a stress on the value-orientation of rugged individualism. On the basis of this value stress a prediction was made that the Homesteaders would reject a proposed plan of contributing their labor to the coöperative effort of building a gymnasium. This is precisely what happened, and long speeches were made to the effect that "I have to look after my own farm."

Two special types of the choice-situation would appear to be particularly fruitful in values research: the *crisis* and the *conflict*. The *crisis-situation* may be defined as one which disrupts (or threatens to disrupt) the system of relationships (man-to-man, man-to-nature, or man-to-universe) in a culture and demands immediate action of some kind if the social group is to continue. This type of situation has not only the merit of requiring action (so that the observer can be reasonably certain of observing choice-behavior within a short period of time), but also the further advantage that certain value-orientations are accentuated during such periods of crisis. For the purposes of values research there would appear to be two basic types. The first is a crisis for which there is already present in the cultural system a ready-made behavioral response. The response involves value-commitments, but the selection is now a matter of history. A clear example of this type of situation in Homestead is the hospital treatment of serious illness. Alternatives are genuinely open in this situation as evidenced by the fact that a few Homesteaders have traveled all the way to Oregon to have illnesses treated by a "faith healer." But by and large, the response to serious illness is to take the patient to the nearest hospital as soon as possible. The second type is a crisis for which there is no precedent in the experience of the group as to what line of action to follow. In this case, the choice made is improvised on the basis of values previously applied to other experiences. An example of this type of situation occurred when one of the Homesteaders "went crazy" and became violent. There had been previous cases of mental disorder in the community, but this was the first instance of violence associated with insanity. For reasons which are understandable in terms of the value system (and which will be discussed later) the services of the deputy sheriff in the community were rejected, and the psychopath was taken away some three hundred miles to a mental hospital by an informally organized group of community leaders.

The concrete types of crises occurring in the round of community life which are fruitful for values study may be classified as follows:

1. Pressures from the natural environment, including drought, heavy snow, windstorms, frost, hail, lightning.
2. Life crises, including birth, death, illness, accidents.
3. Property crises, such as fire or theft.
4. Crises in human relations, such as fights, suicide, or murder.

The second special type of choice-situation is that of the *conflict-situation*, in which one or more value-positions come into conflict. Here again values are accentuated in situations which facilitate observation, although no immediate action is required. The types of conflict situations which were found to be fruitful in the study

INTRODUCTION

of Homestead were cross-cultural marriages; conflicts with Spanish-Americans for the control of land, politics, and the school system; conflicts of factions within Homestead; role conflicts (especially husband-wife and parent-child); and conflicts between homesteaders and ranchers.

Second in importance to observing, interviewing, and giving questionnaires focused upon choice-behavior to samples of informants is the method of content analysis of the daily flow of statements, and especially gossip, which inevitably reach the observer while simply living in a community. It is critically important to catch these statements in the ethnographic record, especially those which are constantly approving or disapproving given courses of action. Over a period of several months these data immensely deepen and enrich one's knowledge of the community's values. This method can be sharpened considerably in its use in values research by focusing upon two sets of persons in the community: (a) those whom the community defines as "bad," and (b) those whom the community defines as particularly "good." A comparison of the similarities and differences in behavior of the two kinds of persons adds precision to the observer's statements about values. It might also be noted that while the study of choice-behavior seems to be appropriately suited to discovering the *selecting* aspects of value-orientations, the content analysis of the observations of the daily flow of life in a community provides rich data on the *regulating* aspects of value-orientations.

Finally, there are a number of techniques by which an observer may tap the *goal-discriminating* aspects of value-orientations. One technique which was used with success in Homestead was that of collecting autobiographies from each student in the high school. The autobiographies were in two parts: a description of the life-career up to the present, and a projection of the life-career to 2000 A.D.[11]

Utilizing these methods, the Homesteaders' major value-orientations can be described as follows:[12] a strong stress upon *individualism* in social relations; an accent upon what I shall call *hopeful mastery over nature* in their relationships to the physical environment; an emphasis upon the *future* as the important time-dimension; a patterned balance between *working and loafing* as the desirable way to allocate activities throughout the year; and a very complex combination of *group-superiority* and *group-inferiority* orientations in their relationships to other cultural groups.[13]

VALUE-ORIENTATIONS AND CULTURE PROCESSES

Precise description of the value-orientations of different cultures will represent an important step in values research. But once we have described and classified these value-orientations, what then? What can we say about the relationships which exist between value phenomena and other determinants of human conduct, and can we determine what effect different systems of value-orientations have upon the course of human events? It is the aim of this inquiry to make some theoretical suggestions about the role of value-orientations in cultural processes and to test these statements against the empirical materials on the Homesteaders.[14]

These theoretical suggestions focus upon the conceptual model outlined in the section on "Definitions and Concepts." The basic postulates in this model are twofold. First, "man-in-culture" is not merely pushed and motivated by economic considerations on the one hand and by biological impulses on the other; rather, man views the world through cultural lenses (which are ground from particular combinations of value-orientations) and reacts in terms of value-patterns which are constantly selecting, regulating, and orienting human behavior in ways that override factors of sheer tension-reduction. Second, the essence of these regularities in behavior must be analyzed with dynamic processual concepts as well as with structural concepts.[15] In other words, cultures are in continual process of change — indeed, it may be argued that the basic tendencies in cultural systems are toward change, rather than toward coming to rest in static equilibrium states — and to use only structural concepts in the analysis of cultural phenomena is like arbitrarily stopping a movie to study the patterns of a single frame instead of studying the patterned sequences of events that make up the total movie.[16] For certain types of problems, it is necessary to treat culture *as if* it were only a static structure and to use primarily structural concepts and categories in analysis, and the development of systematic structural theory has represented an important advance in basic social science. Indeed, structural concepts are utilized at certain important points in the present study. But for the central types of problems under consideration in this inquiry, it is more

INTRODUCTION 13

meaningful to treat culture as a set of processes and to take a series of "readings" on a cultural system over as long a time span as possible for which data is available.[17]

By "processes" I refer to continuous changes in time recurring in orderly sequences of behavior. If a given cultural manifestation is conceptualized as a set of recurring orderly sequences of behavior, the continuous change which occurs is obviously not random but channeled in certain directions by the interaction between situational requirements and value-orientations in their functions as selectors, regulators, and goal-discriminators. It is probable that in a long-run development of culture (over periods of hundreds or thousands of years), the cultural system "shakes down" and either an adjustment is made to the ecological and economic situational realities, or the culture ceases to exist.[18] But in shorter time periods, there would appear to be a dramatic "playing out" of the crucial value-orientations in a manner that overrides the realities of the situation.

In the time period of this study, it can be shown that not only do value-orientations influence the utilization of the natural environment, the modes of social organization, and the direction of change, but that a set of value-orientations may even transcend changes in the environmental and economic aspects of the situation of action in such a way as to contribute significantly to the transformation of a local community.

CENTRAL PROBLEMS

In translating these general ideas into concrete terms in the case of the Homesteaders, I shall advance the thesis that a cluster of the crucial value-orientations had a positive effect on the settlement and early development of the community, but that the continued (and almost compulsive) adherence to aspects of these same value-orientations in the face of a changing environmental and economic situation is now contributing to the transformation of the community. In other words, the situation, in both its environmental and general economic aspects, has altered in recent years, but the value-orientations have not changed proportionately. Further, the values which were appropriate for the original settlement and early survival of the community are now inappropriate for its continuing development.

This suggests a case in which the cultural system is not self-correcting from the point of view of the preservation of the local community.[19] Rather, the process forms a temporally connected sequence of events which is circularly reinforcing in the direction of transforming Homestead into a different type of cultural unit from that which the founders had in mind.

Specifically, the argument is that the strategic value-orientations of "rugged individualism," mastery over and exploitation of the natural environment, and an optimistic faith in future progress were important positive factors in the migration from Texas and Oklahoma and in the initial settlement and early development of Homestead. It is doubtful that the community would have survived at all in this hazardous natural environment if these values had not been strongly present in the cultural tradition. Although through the years the situation has changed markedly in its environmental aspects (drought and depletion of the soil from wind erosion) and in its general economic aspects (especially the rise to economic power of encroaching cattle ranchers), the fundamental stress on these strategic values has not diminished; nor have other, more appropriate value-orientations emerged within the local culture. Indeed, it can be demonstrated that the response to the threatened decline of the community has tended to center around more vigorous expressions of the same strategic values, with a cumulative effect upon the transformation of the community.

These empirical considerations suggest the more general proposition that the combination of rugged individualism with its implications for free market behavior, the stress upon mastery over and exploitation of the natural environment, and a future-time orientation forms necessary value-orientations for the successful settlement and development of this homesteader type of frontier community in American culture, but that unless these values are eventually balanced by at least some stress upon social coöperation and the conservation of natural resources, the ultimate outcome of the cultural process will be self-transformation rather than self-correction and preservation of the frontier community. What in fact will eventually emerge in the Homestead ecological situation is a group of widely scattered ranches and no community center with its attendant service institutions.

But before the value-orientations are discussed in detail, it is necessary to present a brief historical sketch of the community and

INTRODUCTION

the "hard facts" about the geographical, demographic, and economic situation (Part I). I shall then, in Part II, show how the Homesteaders reacted to this situation in cultural and more especially in value terms as they attempted to develop a new community in western New Mexico. Each of the major value-orientations will be treated in turn in successive chapters, except that *individualism*, which pervades so much of Homestead culture, will be discussed in the chapter on "The Atomistic Social Order," where I shall analyze the consequences of the Homesteaders' values for the over-all structure of the community. The final chapter will then deal with the role of these value-orientations in the cultural processes that are manifested in the settlement, development, and current transformation of Homestead.

HISTORICAL SKETCH

Homestead was founded in the early 1930's by families from the South Plains region of western Texas and Oklahoma. This migration to Homestead represented a small part of the vast westward expansion of American people from the older settlements in the East to pioneer settlements on the Western frontier.[20] In particular, the movement to Homestead was part of the essentially *southern* migration from the Atlantic Seaboard to the hill country of Tennessee, Georgia, and Alabama to the Plains of Texas and Oklahoma and thence (later) on to the West Coast. The latest migration from the Plains westward to California was popularized in Steinbeck's *The Grapes of Wrath* and was the subject of investigation by many governmental agencies in the 1930's and 1940's.[21]

During the decade between 1930 and 1940 it has been estimated that approximately 1,200,000 persons migrated to California from states further east. Of this number, one-fourth (or 400,000 persons) were drawn from the states of Texas, Oklahoma, and Arkansas, and the majority of all of the agricultural people migrating to California were drawn from these states. This mass migration started in 1930–1931 and reached a peak in 1936 and 1937, almost half of the migration occurring during these latter two years. The popular impression has tended to identify "the Dust Bowl refugees" with the entire migration to California in this decade. This is clearly not an accurate impression since only a fraction of the migrants were displaced agricultural families from the South Plains. They were, nevertheless, a significant part of the total migration and were the cause of critical economic and social problems in the states affected.[22]

This mass migration westward coincided in time with the period of national depression and with a period of difficult agricultural conditions. On the South Plains there was a good wheat crop in 1931, but prices were low and continued to be low into 1932 and 1933. The great drought, with high velocity winds, struck in 1934 and continued in 1935 and 1936.[23]

In California the new migrants became conspicuous in the state's depression problems, such as relief, unemployment, health costs, and housing. The word "migrant" was used in this decade not as a term of honor describing a latter-day pioneer; rather, it became synonymous with "indigent," with "drought refugees," with habitual "migratory workers," with the Joads of *The Grapes of Wrath*. In sharp contrast with previous decades, popular attitude toward newcomers in the 1930's was unfriendly almost everywhere and actively hostile in many sections of the Pacific region. A widely known example was the "bum blockade" maintained by the Los Angeles city police force during the fall of 1935 at the southeastern border of California. Police stationed at the border turned back or arrested "migrants" bound for Los Angeles who were without means of support.[24] However, the migrants who stayed on in California were eventually given employment as the nation revived from the throes of the depression, and many of them became the backbone of the labor force for the large-scale agricultural enterprises in the great Central Valley of California.[25]

Some of the migrants, instead of going on to California, stopped in New Mexico to settle in a number of semiarid farming areas in the northern and western parts of the state. Some stayed in these communities; others eventually moved on to California; still others returned later to the Plains or resettled in the Rio Grande Valley. Homestead is one of few such communities to survive the vicissitudes of nature, the depression, and the attraction of economic alternatives during World War II and the post-war years.

The information that land was available for homesteading in western New Mexico reached the South Plains through families who had settled in the Pueblo Plateau country north and east of Homestead in an earlier wave of migration in the 1920's. When the depression struck, they wrote letters to relatives and friends back in Texas and

INTRODUCTION 17

Oklahoma and urged them to move west to take up homesteads in this modern "Garden of Eden." So the Pueblo Plateau became the scene of a pioneer movement in the twentieth century. A few of the families came in wagons, with their hogs and chickens aboard, and their milk cows driven along behind. But most of them came in trucks or old cars over the plains of eastern New Mexico, across the Rio Grande and on to the plateau.

The first settlers arrived to look over the open land in 1929, and the next spring they began to file homestead petitions. The usual procedure was to pick out a section and then file an application in Santa Fe. During the two-year period from 1930 to 1932 eighty-one families arrived and filed upon homesteads. Over 90 per cent of the families came from the South Plains country in the Texas Panhandle; the others were from other regions of western Texas and Oklahoma. The South Plains area is on the border zone between the wheat region to the north and the cotton region to the south. Plainview, the ancestral home of many Homestead families, is advertised as the place "Where King Cotton Meets Queen Wheat."

Back on the Plains almost all of these people came from agricultural families. About 10 per cent of them were small landowners with heavy mortgages on their farms; the others were tenants, farm laborers, or surplus sons from small family farms. However, they did not all arrive in Homestead without resources. In Homestead today the observer is likely to hear more about the dramatic cases of poverty than about the families with means. One group of four brothers could not even pay the $34 filing fee on a homestead when they arrived; a widow with her three children came in a covered wagon, and her assets were only $1.49 in cash, a team of horses, eight cows, and a few hogs and chickens; a number of others were able to pay the filing fees on their homesteads but had "to pick pinyons to get by the first winter." But the fact is that 58 per cent of the families had resources of cash, livestock, or equipment which ranged in value from $500 to $2000 or more. These conclusions are in accord with Goodsell's findings about the homesteading movement on the whole Pueblo Plateau. Goodsell writes:

> They were attracted to this area largely by land for homesteading. The settlers' general enthusiasm for their new-found homes has attracted wide attention. These farmers believe that they have settled in an area of promise for farming activities. They did not merely drift into the area

under duress of poverty, as might be presumed from the fact that many of them hailed from the plains during the historic drought of that area. This is evidenced by the fact that upon arrival, 22 per cent of them possessed from $500 to $5,000 and over in cash, and another 15 per cent brought from $100 to $500 with them. More significantly, 55 per cent brought assets in the form of cash, livestock, farm machinery, and automobiles and trucks, the cash equivalent of which ranged from $500 to $2,000 or more. Only 11 per cent had debts which they incurred at their former place of residence. It is evident that here is a genuine pioneer movement, comprised of substantial people in quest of permanent farm homes.[26]

It is clear that while the movement to Homestead was part of the vast migration of southern families westward, the group who migrated to western New Mexico differed from the larger group of 400,000 who passed on through to California in the 1930's in two important respects: (a) most of the homesteaders migrated early in the decade, before the peak of the California migration was reached; (b) the majority of the homesteaders did not migrate under utter duress of poverty but were more substantial people than the Joads and other "Okies" of *The Grapes of Wrath*. While the immediate impetus for the movement was provided by the depression and by the severe agricultural conditions on the Plains, the long-range promise of an opportunity to establish permanent family-owned farms on which they could be "independent" and control their own destinies, rather than being tenants or working for somebody else, was a critical factor in the decision to migrate. Furthermore, while some of the "big ranchers" in the Homestead area actively discouraged and opposed the homesteading effort, the general reception was quite different from that accorded the migrants in California. They were defined more as genuine twentieth-century pioneers; newspaper articles were written about the virtues of the community; a prominent banker in Gallup was (and has continued to be) thoroughly sympathetic toward the Homesteaders in the granting of farm loans; and various governmental agencies came to the support of the community in the early days.

This is not to say that life was easy in the pioneer days of Homestead. The more comfortable and settled villages of the Panhandle had been left behind, and the pioneers faced the usual frontier problems of building homes; clearing land for cultivation; establishing stores, a post office, and a school; and obtaining supplies and

INTRODUCTION

marketing products from a very isolated area. Above all, their pinto bean crops were always threatened by the natural hazards of drought, frost, wind, and hail.[27]

When they arrived in western New Mexico, the new settlers found that the area was populated by scattered communities of different kinds of people, some of whom they had never seen before, except perhaps in the movies. Nearby was the "Mexican" village of Tapala; to the north were two American Indian communities, Pueblo and Navaho, and the Mormon village of Rimrock.[28]

Beyond these local communities of "Mexicans," Indians, and Mormons were the various towns (Gallup, St. Johns, Quemado, Grants, and Albuquerque) which served as the shopping, marketing, medical, and recreational centers for the Homesteaders and were their principal links with the outside world (see sketch map of the Homestead area). The isolation of Homestead even from these towns is great, for the roads are still unpaved and no telephone lines reach the community. In the rainy season and during the winter it is often literally impossible to reach any of the other towns by automobile for days at a time.[29]

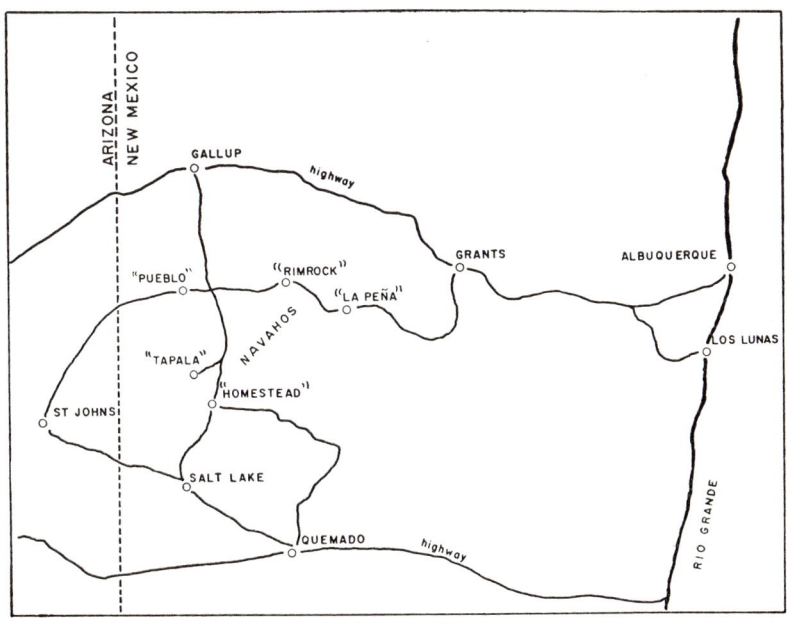

SKETCH MAP OF HOMESTEAD AREA

PART I   THE SITUATION

1

The Geography and Demography of Homestead

When the Homesteaders arrived in western New Mexico they entered a geographical situation that differed markedly from their familiar environment on the South Plains in Texas and Oklahoma.

THE NATURAL LANDSCAPE

The Homestead area is situated in the high mesa country near the southern edge of the Colorado Plateau at an elevation of 7000 feet (± 200 feet), the elevation of Homestead center being 7069 feet. To the south is a high escarpment, known locally as "the rim," which rises about 1000 feet above the Salt Lake valley to the south and marks the southern border of the community. From this escarpment the surface slopes gradually, interrupted by a few rocky knobs, to the north and northeast through the center of the community and reaches its lowest elevation (about 6800 feet) in the extreme northeast corner of the area. Eastward the area rises again over the top of a series of low rocky ridges, running generally north and south. Beyond these ridges there are many ancient basalt flows which limit the amount of arable land on the eastern periphery and form a sharp northern boundary to the community. To the west the high gently rolling tableland breaks off abruptly into a series of deep, rugged canyons.[1]

GEOLOGY AND SOILS

The high escarpment south of Homestead exposes the upper geological formations underlying the area. These formations range from the Quaternary basalt flows down through Tertiary sands and conglomerates of the Pliocene age to the Mesa Verde formation of the Upper Cretaceous age.[2]

The Mesa Verde formation consists of about 1800 feet of alternating gray to buff sandstones, gray clay shale, and coal. It underlies the Tertiary deposits west of Homestead, but outcrops in the escarpment south of Homestead where almost the entire thickness is exposed. Eastward from Homestead the formation is exposed over a rather broad area but is concealed in part by the scattered remnants of Tertiary sands and by basalt flows.

During Tertiary time, the shales and sandstones of the upper Mesa Verde formation were eroded and subsequent deposition of the Tertiary deposits filled the old channels. These Tertiary deposits consist of loose to slightly indurated white sand and a lower, well-cemented, coarse white conglomerate which contains numerous cobbles and boulders of basalt. This Tertiary formation now constitutes the surface formation in most of the western half of the community. The flows of more recent Quaternary basalt cover most of the eastern and northern peripheries of the Homestead area, and scattered remnants are found in the southeast.

Ground water occurs in the Mesa Verde formation (which is the main aquifer of the area) and in the Tertiary formation. However, the existence of buried ridges and channels in the anciently eroded upper member of the Mesa Verde formation and the recent erosion of the Tertiary deposits result in significant variations in the depth from the surface to ground water supplies. Structural conditions of the area are not yet well known, but there are indications of several faults and one syncline which further complicate the ground water situation.

The surface soils produced from these formations are highly susceptible to erosion by wind and water, but they are deep (36 inches or more) and are moderately to well developed, ranging in texture from sandy clay loam to clay. Soil fertility and physical characteristics are such that these soils provide excellent conditions for plant growth.

FLORA AND FAUNA

All of the Homestead area is located within the pinyon-juniper belt, known as the Upper Sonoran Zone, which extends from Colorado and Utah south to central Arizona and New Mexico, and reaches to some extent on the west into eastern Nevada and California and on the east into western Kansas, Oklahoma, and Texas. The altitudinal range of this zone is from about 4500 to 7500 feet and its limits vary with differences in exposure and moisture conditions.

GEOGRAPHY AND DEMOGRAPHY 25

Before the Homesteaders arrived, the area was covered with stands of pinyon and juniper trees.[3] Here and there between the groves of trees, and especially in low depressions which collected water and formed lakes in the rainy season, there were a few treeless areas, and some of these are now in cultivation. Most of the region, however, had to be cleared of its pinyon and juniper trees before it could be farmed. In the deep canyons on the western edge of the community there are intrusions of ponderosa pine and Gambel oak from a higher plant zone, but these regions are not used by the Homesteaders except for hunting.

Sagebrush, which is found so commonly in the pinyon-juniper woodlands of the Southwest, does not grow in the Homestead drainage area, and the most common types of shrubs are rabbitbrush and snakeweed. These shrubs are not used for any purpose by the Homesteaders, but the blue gramagrass which grows everywhere between the shrubs is excellent forage for livestock and formed the plant basis for the cattle- and sheep-ranching enterprises that antedated the arrival of the Homesteaders in this region of New Mexico.

Other common plants filling the natural "niches" of this plant zone include the omnipresent Russian thistle; two varieties of yucca, the broad-leafed and narrow-leafed; the prickly pear and the buckhorn cholla; and many species of flowering shrubs and herbs that are characteristic of the Upper Sonoran Zone and grow profusely in rainy years — the Indian paintbrush, Rocky Mountain beeplant, scarlet bugler, Mariposa lily, larkspur, long-flowered gilia, locoweed, and Pingüe. The larkspur, the locoweed, and the Pingüe are poisonous range plants — the larkspur affecting cattle, the locoweed affecting horses, and the Pingüe affecting sheep as well as horses and cattle.

The fauna of the Homestead area are also typical of the Southwestern Upper Sonoran Life Zone.[4] Mule deer are found in sufficient numbers in the rough canyon country on the western border of the community to furnish an important source of meat for the Homesteaders. Mountain lions are rare, but occasionally enter the region to feed upon the deer. Other predaceous animals, especially the coyote and two species of wild cats, have been common in the area, but are also rare at the present time since the New Mexico Bureau of Biological Survey has carried on a vigorous campaign over the state to eliminate them.

Although the large herds of prong-horned antelope are generally found in the lower, more open country to the south and west, they do graze in small herds in the Homestead region and are occasionally hunted by the Homesteaders.

The smaller forms of mammalian fauna are, however, a much more conspicuous feature. The Texas jackrabbit and the Cedar-Belt cottontail were abundant until they were killed by the "plague" a few years ago. They are now reappearing in small numbers.

The prairie dog population was heavy throughout the region in the early days of settlement, but the efforts of the Biological Survey combined with the recent "plague" have virtually eliminated this rodent from the natural setting. Over a three-year period (1949 to 1952) only two prairie dogs were observed in Homestead.

The Arizona porcupine lives in the pinyon groves and makes frequent summer raids into the Homesteaders' fields of corn. The striped skunk is common and often feeds on chickens. Mexican badgers, cliff chipmunks, rock squirrels, kangaroo rats, and pocket gophers are also observed frequently. Pack rats build nests in the woodlands and gather pinyon nuts, and in good years their nests form an important supply of nuts for pinyon-pickers.

It is unnecessary to list all of the many species of birds that are found in the Homestead area, but a few of the more conspicuous forms may be mentioned. By far the most numerous are the pinyon jays which travel in large flocks and often completely consume a pinyon crop before the nuts can be gathered. Two types of hawks are common: the desert sparrow hawk and the red-tailed hawk, the latter having a notorious reputation as a consumer of the Homesteaders' chickens. There are also two types of owls, the burrowing owl and the western horned owl, which is also a chicken-eater. Golden eagles nest in the cliffs about six miles north of Homestead. Crows and turkey vultures are observed occasionally.

The most conspicuous smaller species include the western robin; the Arkansas kingbird, which chatters incessantly in the pinyon trees all summer and is "mocked" by the western mocking bird; the flocks of western mourning doves which feed contentedly in the Homesteaders' fields; the red-shafted flicker which feeds largely on ants; and the nighthawk or "bull bat" which startles one with its whirring dives in the late afternoon and evening.

The Homestead setting is relatively free of pestiferous insects.

Few mosquitoes are found, but there is a two to three week "gnat season" in June and early July, when swarms of gnats make life unpleasant for the farmer in his fields. Cutworms are also a serious problem, especially in their destruction of the corn crop, in some years. Otherwise insects are not mentioned by the Homesteaders as creating any serious problems.

CLIMATE

The general climatic patterns of the Southwest are now well known for both the "eastern" and "western" zones that are demarcated by the Rocky Mountain chain.[5] Because Homestead is in the western zone, the following discussion will center on the climatic patterns characteristic there and the local variations in the patterns that are found in the immediate Homestead area.

Climatologists find that there are five principal source regions of air masses contributing to the climate of the Southwest: (1) Cold, dry Polar Continental (Canada and northward); (2) Cool, moist Polar Pacific (northern Pacific Ocean); (3) Hot, dry Tropical Continental (Mexico, extreme southwestern United States); (4) Warm, moist Tropical Gulf (Gulf of Mexico and Caribbean); (5) Warm, moist Tropical Pacific (southern Pacific Ocean). Although the weather at any given place may be influenced in many ways by the movement and interaction on these various air masses, somewhat normal directional movements exist during the various seasons.

In the western zone the *summer* type of storm usually develops when moist, warm Tropical Gulf air invades southern Arizona from Old Mexico and moves northward into northern Arizona and New Mexico. Alternatively, the moist, warm air may move into western New Mexico more directly from the Gulf of Mexico, rather than following the more circuitous route through Arizona.

Although Tropical Pacific air does not invade the Southwest except under unusual circumstances, it should be noted that when this type of invasion does occur, there result periods of excessive and widespread rainfall.

During the *winter* months precipitation ordinarily occurs when moist Polar Pacific air invades the western zone from the north, northwest, or west. Alternatively, some winter moisture reaches the western zone of New Mexico in invasions of air from the east or southeast, but this is not as regular an occurrence.

Dry periods result in the *summer* months when Tropical Continental air is developed over the Southwest from the heating of an air mass present in the area that has already lost its original characteristics. The weather during such periods is characterized by high temperatures, low humidity, and an almost complete lack of precipitation. This condition is particularly prevalent in western New Mexico during the late spring and early summer.

During the *winter* months Polar Continental air may invade the Southwest from the north or northeast. Such invasions are often characterized by intense cold, and the extreme dryness of this air mass usually precludes the occurrence of precipitation unless other sources of moisture are available.[6]

Like the rest of the western zone, the Homestead area has two rather distinct rainy seasons, one during January-March, and the other during July-September. The first is due to the activity of the Polar Pacific air, the second due to the Tropical Gulf air. During the transitory dry months of April-June, the activity of the Polar Pacific air is greatly lessened, and the invasions of moist Gulf air have not yet begun. During September-December the invasions of Gulf air are diminishing in intensity and frequency, and the activity of Polar Pacific air is not great. During the summer dry period the weather is controlled by the Tropical Continental air; during the winter dry period there are frequent invasions of the dry Polar Continental air.

With the prevailing elevation of 7000 feet, the winter storms usually come in the form of snow which in some winters piles up to a depth of over two feet. The summer storms typically are scattered thundershowers, but occasionally there are general rains over wide areas. And there is good evidence that the high escarpment which forms the southern boundary of the Homestead area has an effect on the incidence of these summer thundershowers. This escarpment, in providing a sudden change in topography from an elevation of approximately 6200 feet in the valley to the south to an elevation of around 7200 feet on top of the "rim," has an orographic effect upon the northward-moving Gulf air masses and results in a greater summer rainfall for the Homestead community.

But while this discussion of the predominant Southwestern climatic patterns and of the characteristic patterns in the Homestead area helps us to understand the local climate in long-range terms, it

GEOGRAPHY AND DEMOGRAPHY

must be kept in mind that *great variations* may occur from *year to year* and from *place to place* in any given locality. As indicated in Table I (below), the annual rainfall in the Homestead area has varied between six and nineteen inches over a nineteen-year period. Furthermore, there is always variation in any given year within a small locality — summer showers have been known literally to "stop

TABLE I
Annual precipitation and bean production in Homestead [7]

Year	Precipitation *	Yield of beans in 100-lb. sacks per acre	Average price per cwt.
1932	13.60 (Q)	3.0	$2.50
1933	18.95 (Q)	5.5	3.50
1934	14.58 (H)	4.2	4.00
1935	15.14 (H)	4.9	5.00
1936	12.89 (H)	4.7	4.14
1937	12.29 (Q)	4.1	4.00
1938	10.77 (Q)	1.3	3.75
1939	11.58 (EMM)	0.7	3.15
1940	18.91 (EMM)	3.0	2.25
		Average 3.5	
		Total bean production by 100-lb. sacks	
1941	19.01 (EMA)	6000	3.50
1942	9.14 (EMA)	6000	8.00
1943	11.06 (EMA)	6500	5.50
1944	8.20 (EMA)	5500	5.35
1945	7.12 (EMA)	5400	5.25
1946	10.63 (EMA)	13000	12.50
1947	15.19 (EMA)	7000	8.00
1948	12.67 (EMA)	11380	9.00
1949	10.95 (EMM)	8300	7.10
1950	6.00 (EMM)	3300	7.60
Average 12.56 inches		7238	$5.47

* These precipitation data are from the United States Weather Bureau Office in Albuquerque, New Mexico. Reliable figures from the community of Homestead are available only for the years 1934–1936, when a weather station was maintained in the village. The other data are taken from the closest weather station (of comparable elevation) to Homestead for years when the figures were complete for each station. Q is 28 miles SSE of Homestead at an elevation of 6879 feet; EMM is 32 miles NE at an elevation of 7218 feet; and EMA is 28 miles NE at an elevation of 7120 feet.

at a barbed-wire fence" in Homestead. The factors which account for these variations are immensely complicated, and have to do with the fluid interaction of the various air masses, the topographic relief, the changes in elevation, and location with reference to storm movements. It is these local variations which are the common concern of the Homesteaders in attempting to raise crops of pinto beans with dry-land farming techniques.

While the average growing season in Homestead is 110 days, the time between killing frosts in the late spring and the autumn has varied between 82 and 169 days. There is constant danger that the bean crop will be killed by late spring frosts, which have occurred as late as June 18, or damaged by early fall frosts, which have occurred as early as September 8. The mean temperature is 52.7 degrees Fahrenheit. The monthly average temperatures vary from 25 degrees during January to 66.2 in July, but winter temperatures have been known to reach 48 degrees below zero, and summer temperatures have been recorded as high as 116 degrees. However, the extremes are rarely above 95 or below minus-30 degrees.

The heaviest winds in Homestead occur in the spring, especially in the months of March and April. They coincide with the spring dry period and with the period when there are no growing crops or stubble on the bean fields. The result is that the topsoil blows off the land drastically, and these are "dark days" that grimly remind the Homesteaders of the "dust-bowl" period back on the Plains.

MAJOR GEOGRAPHICAL PROBLEMS

Having described the principal features of the natural setting, we must now summarize and assess the relevance of these geographical variables for the agricultural occupation of the Homestead region.[8] It is evident that the Homesteaders brought a new type of subsistence pattern into a region that had previously been utilized only by ranchers grazing livestock on the stands of blue gramagrass, by occasional hunters of deer in the area, and by Indian and Spanish-American pinyon-pickers in good pinyon years. These earlier land-use patterns had a history of fifty years when the Homesteaders arrived and proceeded to implant a dry-farming pattern upon the land.

It was, of course, impossible for the Homesteaders to undertake extensive ranching operations since they began their settlement of the region with only one section of land per family, and one sec-

GEOGRAPHY AND DEMOGRAPHY 31

tion would support only eight to ten cows. The only alternative was to develop dry-farming techniques that would work in this semiarid environment. Supplementing the dry-land crops, a few head of livestock could be grazed on the pasture land on each section, and some use could be made of the wild plant and animal life.

In the development of the agricultural pattern, it is clear that the variation in intensity and in annual distribution of rainfall is the chief limiting factor, with secondary threats being posed by killing frosts and wind erosion.[9] The topography of the land is such that while it is "rougher" country than the Plains country back in Texas and Oklahoma, there is still ample flat and rolling land to make large fields possible. The geological formations have produced deep and fertile soils which — after they are cleared of the pinyon-juniper woodland — provide excellent conditions for field crops.

Since natural lakes and springs are generally lacking in this arid land, the successful development of the water wells for household and livestock use is also necessary. Here the geological structure of the Homestead region poses a problem in survival. Ground water is found in both the Mesa Verde and the Tertiary formations but the distance to water may vary from 88 to over 500 feet in different parts of the community.

Compared to the water problems, the factors relevant to survival and development of the community in the native plant and animal life are relatively minor. But a few important adjustments and relationships must be discussed. The plowing up of the grassland, along with the clearing off of the pinyon-juniper woodland, has created special problems in the long-range use and conservation of the land. The cultivated areas are exposed to serious wind erosion, and when fields are abandoned they do not readily grow back into gramagrass but produce Russian thistle and other weeds that are not so nutritious for livestock. The use of the remaining woodlands for fence posts, firewood, cabins, and the gathering of pinyon nuts continues at the present time, as well as the use of grass for livestock on the remaining pasture lands. However, little or no use is made of other wild plants. A few "greens" are picked for food in the summer, but the many uses of the plant life by Indians in the region for "medicines" and food are generally lacking among the Homesteaders — the yucca and prickly-pear fruit are not eaten, the yucca roots are not used for soap, etc.[10]

The meat of the deer, the cottontail, and occasionally an antelope, contributes to the food supply of the Homesteaders. In the first years of settlement, many of them trapped coyotes, wild cats, badgers, and skunks. But other forms of animal and bird life have either a negative or merely neutral value. Rabbits and prairie dogs have often done substantial damage to crops, and the Homesteaders like to have populations of coyotes and wild cats in the area to control these rodents. On this problem, the Homesteader is in conflict with the sheep-rancher, who insists that the Biological Survey field men eliminate the coyotes and cats to prevent sheep losses.

Although there is some recognition that the "balance of nature" should not be upset and the Homesteaders realize that many of the smaller animals and the birds help to maintain this balance, there is a strong tendency to regard most of these faunal forms as "pests" and to eliminate them when the occasion arises. Skunks, hawks, and owls occasionally feed on chickens and are killed whenever they are seen. Other birds are killed for "sport" by younger boys, but few inroads have been made upon the more numerous species such as the pinyon jay, mourning dove, kingbird, or night hawk.

DEMOGRAPHY OF HOMESTEAD

After the first two years of settlement, additional families continued to arrive in Homestead, but since the available homesteading land had all been taken, the volume of migration dropped off markedly (see chart on immigration to and emigration from Homestead).[11]

Although there has been an average of three families per year arriving in Homestead since 1932, there are obvious fluctuations in different periods during the past two decades. The 1933–1934 arrivals were mainly families who waited for others to relinquish their claims or who "squatted" on land that had been withdrawn from homesteading by the Taylor Grazing Act and stayed on for a time with the hope that the land would be opened for homesteading again. Most of these families eventually left the community.

A large proportion of the families arriving in 1935 to 1938 were "business" people who operated stores, cafes, and other enterprises that were supported by the expanding economy of the village. The period of 1939 through 1944 was marked by almost no additions to the population. Two of the three arrivals in 1942 were "draft dodgers" who lived in Homestead for a short time in an effort to

GEOGRAPHY AND DEMOGRAPHY

Immigration to and emigration from Homestead by families

Year	Arrivals	Departures
1932	81	3
1933	9	3
1934	6	7
1935	7	4
1936	7	13
1937	9	10
1938	6	9
1939	1	6
1940		8
1941		9
1942	3	6
1943		2
1944	1	6
1945	2	2
1946	3	1
1947	3	2
1948	1	1
1949		2
1950		1
1951		12

evade arrest. The arrivals during 1945 to 1947 were mainly war veterans, returning to Homestead after World War II to try their luck at farming and to take advantage of a veterans' farm-training program.

From the first years of settlement there were a few families each year who became discouraged with pioneering or who saw more attractive opportunities elsewhere and left Homestead (see Table II). Again, although there has been an average of five families per year leaving Homestead since 1932, there are definite fluctuations in different periods. A few families left during the depression period of 1932 to 1935, but the peak of the emigration in the 1930's occurred during 1936 to 1939 at a time when it was clear that the government authorities were not going to reconsider and open additional land for homesteading, and when there were two years of crop failures in Homestead. There was again a relatively high rate of emigration to defense jobs during World War II. It should be noted that the families leaving in the late 1930's were smaller than those leaving in the early 1940's, a fact which accounts for a greater drop in the total population during World War II. In the years since the war the emigration was negligible until 1951, which was a year of complete crop failure following a year of near failure.

It is significant to note that since 1935 the emigration has exceeded immigration in Homestead during almost every year. This fact is reflected in the total population figures, which are as follows:

$$\begin{array}{l} 1935-375 \\ 1940-333 \\ 1945-239 \\ 1950-232 \end{array}$$

Each year Homestead also exports young people. Since 1937–1938 when the Homestead high school opened, there have been fifty graduates. Of this number, only five have stayed in the community and forty-five have left. Because the community is small and all the available land is being farmed, there are few job opportunities either on the farms or in the service center for young people out of high school.

Although these figures might seem to indicate a high degree of instability in population, the fact is that fifty-one of the sixty-one families now residing in Homestead have been in the community at least since 1938, and twenty-five of the present families are part of the original group which arrived in 1930–1932. The present Homestead families listed by date of their arrival in the community are as follows:

Date of arrival	Number of families
1930–1932	25
1932–1934	10
1934–1936	8
1936–1938	8
1944–1946	5
1946–1948	5
TOTAL	61

These facts indicate that the great majority of families have stayed in Homestead for over fourteen years, and that the people who came and left again were part of a "floating fringe."

In 1950, sixty-five persons lived in Homestead center; the other

GEOGRAPHY AND DEMOGRAPHY

167 persons lived on scattered farmsteads extending as far as twenty miles from the center of the community. The men numbered 124; the women 108. The average size of family, in a tabulation which defined "family" as any independent household and counted the members actually living in Homestead in 1950, was 3.8; the numerical composition of families was as follows:

Number of families	Number of individuals in family
6	1
14	2
10	3
9	4
11	5
5	6
3	7
1	8
2	9
TOTAL 61	

However, the average size of family in which reproduction has been or is possible was found to be 5.3, in a tabulation which included spouses and children who have died or left Homestead. This indi-

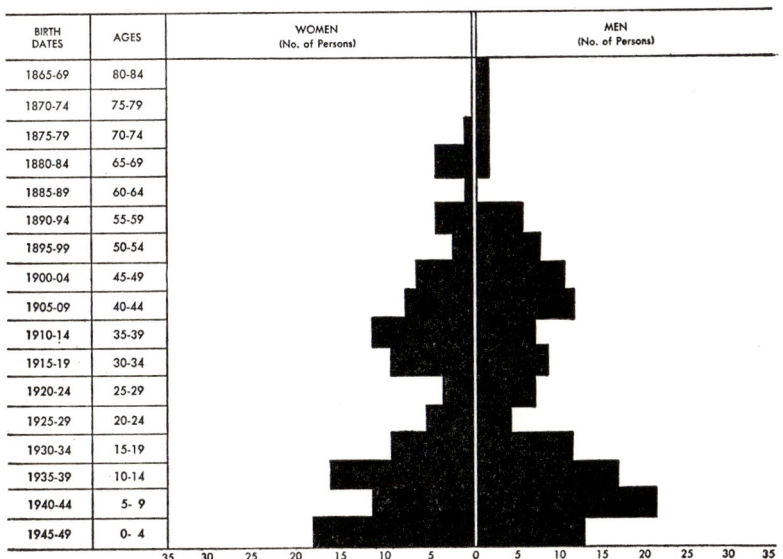

AGE AND SEX DISTRIBUTION IN HOMESTEAD

cates that the total number of children produced by the Homesteaders is somewhat greater than the average size of families presently in the community would indicate.

An age and sex distribution of the population is given on the accompanying chart. Note the relatively few older people in the population, and the fact that there are many persons in the thirty to sixty age-grades, and many in the one to twenty age-grades, with relatively few in the intermediate twenty to thirty range. In general, the thirty to sixty age-grades are composed of active farm operators, many of whom are the original settlers of Homestead, and the one to twenty age-grades are either their children who are still in school or have not left the community, or the second generation produced by those of their older children who have remained. The relative paucity in the twenty to thirty age-grade can be explained by the fact that many of the original settlers do not have children this old yet; or in cases where they do, the children have emigrated from Homestead.

2

The Economy of Homestead

Although the Homesteaders originally settled on the land primarily under the provisions of the 1916 Stock Raising Homestead Act, the economy of the community revolves around the production of the pinto bean (Phaseolus vulgaris). Two decades of experience in this arid land have only served to confirm the Homesteaders' convictions that this hardy plant is the most productive crop which can be grown under dry-farming conditions. The purpose of this chapter is to provide a precise description of the economy of Homestead with its focus upon the pinto bean, but with secondary emphases upon crops of corn and wheat, upon livestock raising, and upon various "jobs" and "businesses" in the town.

THE ACQUISITION OF LAND

The property system in the Homestead area has provided a number of methods by which an individual could acquire land: homesteading, "buying a relinquishment," making an outright purchase, "trading with the railroad," "leasing a school section," acquiring a Taylor Grazing Act permit, and "renting" from a land owner.

Until the Homesteaders arrived in the early 1930's, most of the land which now supports the community was public domain and was used simply for grazing livestock (without formal ownership) by Anglo and Spanish-American ranchers in the area. The act of December 29, 1916 (39 Stat. 862; U.S.C., title 43, secs. 291 and following), commonly known as the "Stock Raising Homestead Act," had provided the legal means by which this public domain could be homesteaded by individual settlers. In the period immediately following World War I and through the 1920's, a number of families appeared and homesteaded land in the vicinity of La Peña, some thirty miles to the northeast, where bean farming continues as an important activity. Others took up homesteads to the east and southeast of Homestead, but many of them later moved away, and this early settlement did not result in sustained agricultural development. It was not until the two-year period of 1930–1932 that settlers began to pour into the region that now comprises the community of Homestead.

Under the provisions of the 1916 Homestead Act, an individual could acquire title to 640 acres (a "section") of land by paying a "filing fee" of $34, living on the land for at least seven months a year for three years, building a "habitable" home, making $800 worth of improvements, and paying a "proving up" fee of $34. It was by means of this law that most of the original settlers of Homestead acquired title to a section of land each.

By the end of 1932 almost all of the available land had been homesteaded and new arrivals found themselves in a position of having to buy "relinquishments." The method was to find a person who had filed on a homestead, made some of the required improvements, and then decided to leave the community. The new arrival would reimburse the original homesteader for the improvements and the latter would then relinquish his claim, leaving the land available to be filed upon by the new settler.

As soon as a homestead was patented and the owner had a formal deed to the land, he was free to sell or dispose of it as he chose. Through the years there has been much buying and selling of land on the open market in Homestead. This method of outright purchase has been the most common means by which original settlers have added sections of land to their holdings.

In 1866 Congress passed the so-called Enabling Act, in accord with which railroads in certain parts of the United States were granted public lands in alternate sections extending forty miles on either side of the railroad tracks. This act was later amended so that additional land on both sides of the track was granted to compensate the railroads for losses sustained due to the fact that in some instances settlers or Indians on reservations had prior claims. The Atchison, Topeka, and Santa Fe Railroad participated in this arrangement and acquired large blocks of land, some of which were in and near the region which now comprises the community of Homestead. In a few cases settlers obtained their land by purchase from the Atchison, Topeka, and Santa Fe or by trading sections, which they had homesteaded and patented, to this railroad company for

ECONOMY OF HOMESTEAD 39

other sections which were closer to the center of the community and more suitable for cultivation.

Still other settlers in the community have leased sections of land which the state of New Mexico had selected for the support of public schools. These lands could not be purchased except by being placed on the market and sold to the highest bidder, but the lessee had a measure of protection in that he had first choice to purchase the land if he met the highest bid. If he did not choose to purchase the land, he still had to be paid a fair price by the purchaser for the house and other improvements he may have made on the land during the period that he had leased the section.

The act of June 8, 1934 (48 Stat. 1269), commonly known as the "Taylor Grazing Act," and the resulting Executive Orders, withdrew all public domain in this area pending classification according to its best use in light of public interest. At no time, however, was acquisition of public land prohibited. Consequently, settlers continued to arrive, and many of them simply "squatted" on the land, pending its classification as farming or grazing land. Subsequently, most of the land on which the "squatters" settled in 1934–1935 was classified as unsuitable for farming, and with the exception of one farmer (who steadfastly refuses to be removed from the area), the "squatters" relinquished their claims and left the community or turned to other economic pursuits.

The Taylor Grazing Act provided for the establishment of grazing districts within which the Secretary of the Interior was authorized to issue permits:

. . . to graze livestock on such grazing districts to such bona fide settlers, residents, and other stock owners as under his rules and regulations are entitled to participate in the use of the range, upon payment annually of reasonable fees in each case to be fixed or determined from time to time . . . Preference shall be given in the issuance of grazing permits to those within or near a district who are landowners engaged in the livestock business, bona fide occupants or settlers, or owners of water or water rights, as may be necessary to permit the proper use of lands, water or water rights owned, occupied, or leased by them, except that until July 1, 1935, no preference shall be given in the issuance of such permits to any such owner, occupant, or settler whose rights were acquired between January 1, 1934 and December 31, 1934 . . .

The result of these provisions of the Taylor Grazing Act was that most of the grazing permits were issued to livestock ranchers whose operations in the area and whose control of water rights antedated the arrival of the Homesteaders. They clearly ruled out the claims of the "squatters" who arrived during 1934. Later a few of the Homesteaders who had acquired small herds of cattle were able to obtain grazing permits for a total of some ten sections of land.[1]

The Taylor Grazing Act further provided:

> That the Secretary is authorized, in his discretion, to examine and classify any lands within such grazing districts which are more valuable and suitable for the production of agricultural crops than native grasses and forage plants, and to open such lands to homestead entry in tracts not exceeding three hundred and twenty acres in area.

But, by and large, the immediate area has continued to be defined as "grazing land" by the Grazing Service, and it has been impossible for additional Homesteaders to acquire land.

The latecomers to the community have had no choice but to rent land if such land were available and if they wished to farm a "place" at all. The renting of a "place" involves the following economic arrangements: the owner provides the land and the farmstead, including whatever house and corrals are located on the land; the renter lives on the "place," farms the land with his own machinery, and grazes his livestock on the land. At the end of each farming season the owner receives one-fourth of the agricultural crops produced during the year; the balance belongs to the renter.

PRESENT LAND HOLDINGS

The land base now controlled by the Homesteaders comprises approximately 63,260 acres or 98.8 sections. For the most part these sections are contiguous and are located in three adjacent townships, but some members of the community own sections located as far as twenty miles east of Homestead center, and the eastern boundary of the community is less definite than the sharp boundaries in other directions. On the south the community is bounded by the northern edge of a 150-section cattle ranch. On the northeast the boundary is formed by Spanish-American land holdings (consisting of small ranches ranging in size from two to twenty-five sections); on the northwest the boundary is formed by the southern edge of a 180-section cattle ranch. To the west the community also has a sharp boundary in the form of deep, inaccessible canyons.

ECONOMY OF HOMESTEAD

The Homesteaders' land base has been reduced drastically in size from the 1935 peak in land holdings and population, when approximately 130 sections of land were controlled and the population of Homestead was approximately 375. Of the thirty sections which the Homesteaders "lost" in the interval between 1935 and 1950, twenty sections were purchased by ranchers (principally the ranchers who are now located to the south and to the northwest of the community), and ten sections consisted of land on which prospective homesteaders "squatted" until the land was classified as unsuitable for farming and was returned to the public domain. Eventually grazing permits under the Taylor Grazing Act were issued to ranchers for the use of these ten sections.

The marked decrease in population from 375 in 1935 to 232 in 1950 was not by any means entirely a matter of the Homesteaders losing control of land to the ranchers, because at the same time that the ranchers were increasing their holdings in the area, the Homesteaders who remained in the community were also increasing theirs. Indeed, some fifty sections have been involved in transactions in which the Homesteaders themselves purchased, traded for, or leased land that was originally owned by other Homesteaders who have now left the community. Today the average size of the forty-eight farming units remaining in the community is two sections rather than one section, the size of the original homesteads. In other words, the Homesteaders who remained have, on the average, doubled their holdings since the period of original settlement. This is an economic fact which the Homesteader often fails to realize when he expresses bitterness about "the big ranchers taking over all the land." It is true, of course, that many farming families have been involved in the addition of the fifty sections to their holdings, whereas only two ranching families have been involved in the addition of thirty sections to the area of their ranches. The enormous differential in land ownership between the 150-section and 180-section ranches controlled in each case by one family and the two-section farms controlled by Homestead families also adds to the feeling of economic insecurity and deep resentment on the part of the Homesteaders, especially since each ranch covers substantially more area than the total land base of the Homesteader community.

It is clear, however, that the result of both processes — increase in the ranchers' and in individual Homesteaders' holdings — is a decrease in the total population of Homestead. For when additional land is acquired by either rancher or Homesteader, it does not become a home for another family but is merely added to the existing operating unit, whether this be a ranch or farm or combination.

Furthermore, the tendency to acquire additional land whenever the opportunity arises is almost as marked among individual farmers in Homestead as it is among the ranchers. Indeed, the crucial difference is that the ranchers' occupation of the region antedated that of the Homesteaders and that they had a head start in the acquisition of land. This phenomenon is clearly indicated by the following facts on sizes of farming units in Homestead.

Approximate number of sections in farming unit	*Number of farms*
Less than 0.5	5
0.5	6
1	10
1.5	7
2	12
3	1
4	2
5	2
7	1
8	1
18	1
TOTAL	48

These data on sizes of farms indicate that while almost one-fourth of Homestead's farms are of less than one-section size, over one-half of the farms are now larger than one section, the most common size of unit being in the one- to two-section range. The farm units of less than one-section size include renters who have rented part of an owner's farm; farm owners who have been less successful economically and have sold part of their original homesteads; or farm owners who have recently purchased partial sections of land. The units of three sections or more are owned (or leased) by Homesteaders who have been conspicuously successful economically and have been able to add many sections to their original homesteads. The smaller land owners with one section or less tend to be full-time farmers; the larger land owners with three sections or more tend to farm less and to become livestock ranchers; and the farm

ECONOMY OF HOMESTEAD

operators with one and one-half to two sections combine agricultural crops with livestock.

It is in the group of Homesteaders who control three or more sections of land that one finds the strong tendencies to acquire more and more land and to become more like the old-time ranchers in the area. And with some exceptions (such as the Homesteader who has maintained a bank account of $50,000 for some years, owns less than one and one-half sections, and much prefers farming to ranching) these aspirations are found among most of the smaller land owners in the community who have either been unable to purchase land or unable to find land for sale in recent years. By 1949–1950 the land situation was extremely tight, and it was difficult for prospective renters to find land for rent, much less for sale. When sections did become available for sale, they were usually purchased by the large ranchers. During the two-year period from 1949 to 1951, the two large ranchers in the area purchased four and one-half sections from Homesteader families who were leaving or had left the community.

Homestead is predominantly a community of family-farm owners.[2] Of the 98.8 sections of land now controlled by Homesteaders, only 15.5 sections (or 14 per cent) are rented by tenants. The remaining 83.3 (or 86 per cent) are owned by the farm operators. The number of operators owning and renting land is as follows.

Farm operators who operate only their own land	34
Farm operators who are part-owners, part-renters	6
Farm operators who are renters only	8
TOTAL	48

It is also important to record that six of the total of fourteen renters rent all or part of their farms from close relatives (mother, father, sister, or uncle), and that in only six cases do farmers rent from landowners who do not live in the community.[3] The amount of land rented from these absentee-owners is 4470 acres — or less than 8 per cent of the total acreage.

THE AGRICULTURAL SYSTEM

The present agricultural system is based upon the dry-land cultivation of *pinto beans, corn,* and *winter wheat,* supplemented by *dairy cattle, beef cattle, hogs,* and *chickens.* In addition, approximately one-third of the farms have vegetable gardens irrigated from wells.

The total acreage under cultivation (in the summer of 1950) was 11,213 acres, or 18 per cent of the total land controlled by the community. The acreage devoted to the various crops was as follows.

Crop	Acres	Per cent of total cultivated land
Beans	6417	57.2
Corn	2092	18.6
Wheat	742	6.6
Other crops or fallow	1962	17.6
TOTAL	11,213	100.0

Dairying is a minor economic activity with less than six farm operators selling cream on the market, but most farms keep a milk cow for the home production of milk, cream, and butter. In 1950 there were approximately 1100 beef cattle (predominantly Herefords) on Homestead farms, but over 800 head were concentrated on eight of the larger farm units. Although there were approximately 280 head of hogs in the community in 1950, only two farm operators were in the "hog business" by planting mainly corn and sorghum and feeding these crops to their hogs. But, again, most farm operators keep a few hogs to produce pork for home consumption. One farmer is in the "chicken business" and sells eggs on the market; the others keep a few chickens for eggs and meat for the family.

In short, while the economic activities of the production of dairy and beef cattle, and of the raising of hogs and poultry, are emphasized on a few farms in Homestead, the more typical pattern is a focus upon pinto beans, with some acreage devoted to corn and winter wheat, and with a milk cow, a few hogs, and a few chickens for home consumption only. The land which is not devoted to cultivation is either pasture for the livestock or woodland.

Four major uses are made of the pinyon and juniper timber in the woodland areas of the farms. The early houses were constructed of pinyon logs, and corrals are still made with pinyon or juniper poles. Wood for heating and cooking purposes is cut locally: juniper for "cook stoves" and pinyon for fireplaces and "heaters." The fences are built using juniper posts, and in years when there is a pinyon nut crop, many Homesteaders engage in picking and marketing the nuts.

Twentieth-Century Homesteader Farms Beans with Modern Tractor in an Arid Land

Incomplete Gymnasium Stands as a Monument to Rugged Individualism

The Village Store

ECONOMY OF HOMESTEAD

Most of Homestead's bean crop is marketed through the local bean warehouse. The warehouse operator provides sacks for the beans and then hauls them to the warehouse at a cost of $0.03 a sack. They are run through a bean cleaner which removes the small rocks, dirt, and other extraneous matter that was not removed by the combine. The cost of cleaning and storage was $0.92 a cwt. in 1949.

Once the beans are safely stored in the warehouse, the farmer has two marketing alternatives. He may choose to sell on the open market when he thinks the price is good or when he must sell to meet his debts. Or, if the farmer has previously made the necessary arrangements with the Production and Marketing Administration [*] to receive the support price ($7.10 a cwt. in 1949) for his beans, he may take a "warehouse receipt" to the PMA office and obtain authorization for a loan up to the amount of the support price value of his beans in storage. This step enables him to receive a cash payment for his crop with no delay.

Later in the year the farmer may decide that he can obtain more for his beans than the amount of the support price, and he is then permitted to repay his PMA loan, plus 3 per cent interest and 3 cents per sack for additional storage charges. He may then remove his beans from the warehouse and sell them on the open market. But if he decides that the support price is the best deal he can make, he leaves his beans in the warehouse, and when the loan matures the PMA acquires ownership of the beans.

However, in order to take advantage of the support price and the PMA loans, the farmer must have agreed to limit his bean acreage according to an allotment which he is assigned at the beginning of the farming season. The county in which Homestead is located currently has an optional plan whereby a farmer may choose whether or not to limit his bean acreage and participate in the PMA loan arrangements. The allotments are based upon bean acreage on each farm during previous years, and the field workers from the PMA Office actually measure the acreages on each farm after the crops are planted. Allotments for "dry edible beans" were in effect in 1950, but were not in effect in 1951 because of the emphasis upon increased production.

If the PMA acquires the beans, they are disposed of through various outlets including commercial sales, sales to foreign govern-

[*] Hereinafter referred to as "PMA."

ments, and sales directly to farmers. If they are sold on the market (or disposed of through commercial outlets by the Commodity Credit Corporation of the PMA), many of the Homestead pinto beans find their way to canneries in Texas and other states and some eventually appear on the shelves of Homestead's stores as cans of "Ranch Style" beans.

In years when the crop fails and Homestead's farmers are unable to meet their debts, the "off-season" becomes the time to seek wage-work outside the community. Each year since the settlement of Homestead, a few families have had to seek outside employment during the winter. The numbers of persons involved have depended primarily upon the previous summer's agricultural conditions. After the severe drought of 1950 which resulted in a near crop failure, fifteen families (or one-fourth of the total population of the community) left during the winter to engage in wagework. They worked on road or house construction jobs in nearby communities in New Mexico, or as agricultural laborers in Texas, Arizona, and California. Their gross earnings during the winter varied from $75 to $125 a week and averaged $80 a week.

In this situation of economic uncertainty stemming from drought and frost conditions, it is obvious that elaborate credit arrangements have been necessary for the continuing operation of most farm units. The systems of credit have involved three basic sources: (*a*) government loans from various agencies of the Department of Agriculture and, since World War II, "GI" loans for the veterans; (*b*) loans from banks in nearby towns; (*c*) credit at the local stores and shops and credit at various places of business in the shopping towns near the community. At the present time, the most important sources of credit are the banks, which charge from 8 to 10 per cent interest on loans. The usual practice is to loan up to $500 on a personal note. If more is required, then the farmer must mortgage his crop or his livestock; if a larger loan is needed, the banker asks for a mortgage on the land as collateral.

No figures on the total indebtedness of the community are available, but it is known that over one-half of the farm operators in Homestead have bank loans ranging in amount from approximately $500 to $15,000 at the present time. Thirteen of these loans are from the same bank in Gallup.

ECONOMY OF HOMESTEAD

THE FARMSTEAD

The physical set-up of the farmstead consists of a house, a well, a cellar, a woodpile, a poultry house, a corral and pens for the livestock, and various sheds. The original houses in the community were constructed of pinyon cr pine logs, and approximately one-third of Homestead's families are still living in the same or remodeled versions of these houses. The other two-thirds of the community's families are now living in more "modern" houses constructed of adobes, cinder blocks, "ammunition boxes," or sawed lumber. Adobes, made from the local soils mixed with manure or straw, are the most popular and least expensive form of building material at the present time. Cinder blocks made of volcanic cinders, sand and cement, must be purchased at a cinder-block plant sixty-five miles away. The "ammunition boxes" were acquired during the war at an ordnance depot near Gallup and were used intact, piled one on top of another, providing some measure of insulation. Lumber is purchased at lumberyards in nearby towns or at sawmills in the mountains.

Houses are constructed by the farmer himself with the aid of neighbors and kinsmen who are sometimes hired for pay by the day, and sometimes work to repay other services in the system of exchange labor. All of the skills for building a modern house (adobe-making, masonry, carpentry, plastering, painting, and plumbing) are present in the community, except those for wiring the house for electricity. For this latter task an electrician from a nearby town is employed.

When Homestead was first settled, water had to be hauled in barrels from a lake three miles away, or from a spring located seven miles from the center of the community. Since these two sources supplied the only available water for both household and livestock use, water-hauling was a major task. In 1934 a community well was drilled in the center of town with the aid of government funds. For a period the Homesteaders all hauled water from this central well. Then through the years individual wells were drilled on various farmsteads, and today thirty-three of the forty-eight farm units have wells on the land. A number of these wells have been drilled with funds from the PMA. These wells could not be drilled at "headquarters" but, according to PMA regulations, had to be drilled in another part of the farm. These PMA regulations were designed to encourage greater development of the water resources of the area. But it meant that the farmer who wanted a well at his house site had to bear the expense himself.

Homestead's wells are drilled by one of the two local well drillers at a cost of $3.00 a foot for a six-inch hole. These drillers are farmers who supplement their incomes by drilling wells in Homestead and in the neighboring communities.

The water is pumped from the wells with windmills, and a few farmers now have electric pumps as supplementary power. Stock tanks are put in to water the livestock, and except for the few houses with running water, household water is carried from the windmill in buckets.

Most farms have underground cellars near the house for the storage of home-canned foods and vegetables such as squash, pumpkins, and cabbage. A woodpile with both pinyon and juniper wood is always present. A few farmers now saw wood with power saws, but the usual practice is to cut wood daily with an axe.

Poultry houses are constructed of lumber or adobes; hog pens, of poles or rough lumber. The corrals for the cows are more carefully constructed but are also built of poles or rough lumber. They usually contain one or more milking stalls, calf pens, and perhaps a small area that is roofed over with poles topped with old hay. Only one farmer has a barn of the type that is found in Middle Western or Eastern farming areas.[4] Finally, the farmstead usually includes one or more crude sheds used for wash houses, or for storage of grain, farm machinery, or automobiles.

FOOD

Although pinto beans are frequently eaten (both as green snap beans, fresh or home-canned, and as cooked dry beans), they are no longer the only staple food in the homesteaders' diet. Potatoes, black-eyed peas, corn, and cabbage appear often on the daily menu, either with or without beans. Pork, bacon, and chicken are the common meats. Other common items at the table include cereals, eggs, milk, cottage cheese, buttermilk, coffee, Kool-ade, butter, crackers, biscuits and gravy, syrup, and home-canned jellies, pickles, and fruits.

Less common items of diet are beef (mutton is almost never eaten), venison, rabbit, and fish, and "store-boughten" groceries such

ECONOMY OF HOMESTEAD

as bread, commercially canned fruits and vegetables, and ice cream. For the most part, fresh vegetables are eaten only during the summer and fall when the gardens are producing. During the rest of the year the families eat fresh vegetables at rare intervals when the local stores have a small supply, or when the family buys them on trips to the larger towns. Canning (and more recently freezing) of vegetables and some fruits is necessary to provide food for the family at the least expense. Fruits such as peaches, pears, plums, and apples and some vegetables are purchased from itinerant produce peddlers at the farm door to be canned, preserved, or merely stored in the cellar for winter meals.

Of the most common foods, pork, chicken, beans, corn, eggs, milk, and milk products (buttermilk, cottage cheese, and butter) are all produced on the farm. The potatoes, peas, bacon, flour, cereals, coffee, Kool-ade, and sugar are purchased at stores.

It is customary to eat three meals a day; breakfast in the morning, the heavy meal called "dinner" at noon, and a lighter meal called "supper" at night. Breakfast usually consists of cereal, eggs, hot biscuits, and coffee with cream and sugar. The heavy meal at noon commonly consists of potatoes, beans (or peas or corn), biscuits and gravy, and milk — with the addition of beef, pork, or chicken when they are available from the farm livestock supply or when the family can afford to buy them. The light supper may either be a lighter version of dinner (or leftovers from dinner), or may consist simply of cornbread and milk, or chile and crackers, eggs with biscuits and gravy, or even just cold cereals. Desserts are not considered necessary for everyday family meals. When company is present, however, there is usually pie, cake, ice cream, or cookies. At other times the family may be served jello, home-canned fruits or fruit cobblers, or unfrosted cake. Homemade ice cream is a common dessert for families who have two or more milk cows and a refrigerator.

The pattern of three meals a day varies from family to family. In some cases a mid or late afternoon snack of fresh rolls, cookies, or even homemade fudge or divinity may substitute for the evening supper.

CLOTHING

The prevailing dress of the men for everyday wear consists of "Levis" or khaki trousers, blue or khaki work shirt, work shoes or cowboy boots, and either a narrow-brimmed or broadbrimmed Western felt hat, or (in summer) a broadbrimmed straw hat. Some men wear baseball caps or mechanics' caps. A few of the older farmers wear blue denim bibbed overalls, but these are not common. For funerals, important dances or school meetings, and (sometimes) for trips to larger cities, many of the men wear suits. A few men also wear suits to church on Sunday, but there are months at a time when suits are not worn at all. More common now for dress-up occasions are gabardine trousers with broadcloth or gabardine western-style shirts which are worn to dances and to town, especially by the younger men. In winter many men wear hunting caps and heavy jackets for everyday.

The most common apparel for women are home-sewn dresses made from flowered or printed feed-sack materials or inexpensive cotton house dresses purchased through mail-order houses or at the city stores. The young married women often wear blue jeans, both when doing housework and helping with outside chores, and especially when working in the fields at harvest time. With jeans they wear cotton blouses or shirts, or cotton jerseys in the summer. In the winter a sweater, or a blouse with sweater or jacket, is worn with blue jeans or a skirt. For dressier occasions, such as going to church or to a dance or a trip to town, the clothes worn are similar to those worn by their city cousins. Homestead women are interested in the latest styles they see in the city stores or in the women's magazines, and they put their sewing machines to good use in making dresses, blouses, and skirts for themselves or their daughters. The usual coat for winter is a basic cloth coat. Hats are rarely worn except by a few women at church, but scarves are worn commonly by girls and women alike, especially during cold weather. The older women wear hose for warmth in winter and to the village store. The younger women and girls usually wear anklets and low-heeled shoes for every day, nylon hose and high-heeled pumps or sandals for more formal occasions. Long evening dresses are seldom worn, except by high-school girls at the annual junior prom.

Blue jeans could be called the uniform for boys and girls of school age. Cotton jerseys, broadcloth western-style shirts, or plaid flannel shirts are combined with them. Matching denim jackets are also popular, as well as practical. The girls sometimes wear dresses or skirts and sweaters to school but seem to prefer jeans.

ECONOMY OF HOMESTEAD 51

Pre-school boys are also dressed most of the time in blue jeans and for some special events in corduroy overalls. Little girls are usually to be seen in dresses, although for play around the home they often wear blue jeans or in the winter corduroy overalls for warmth.

HUNTING, FISHING, AND COLLECTING

There are three minor economic activities that are defined as "sports" or "recreation" by most Homesteaders, but which add to the subsistence income: hunting, fishing, and the collecting of pinyon nuts.[5] Of the three, hunting is the most important. The two animals hunted for food are deer and rabbits. Rabbits are a very minor item of diet now, but venison continues to form an important source of meat supply. An average-sized buck can be butchered and put into the deep freeze and will provide meat for a family for over a month. No exact count of the number of deer killed by the Homesteaders in a year is available, but it is known that the majority of the families in the community eat at least one buck a year. In good hunting years, probably 10 per cent of the total meat supply is venison.

Fishing (for which one has to travel to lakes or rivers 50 to 150 miles away) is less popular than hunting, but three families add significantly to their diets by frequent fishing trips.

The collecting of pinyon nuts is a sporadic activity because crops are produced on the average of only once every four or five years. During the first year of settlement, many Homesteaders collected enough pinyons to buy groceries for the winter. Today, although most families go out to collect pinyons for home consumption, a few individuals go into the "pinyon business" by leasing the pinyon woods on several sections of land and importing crews of Navahos to collect the nuts. In 1948 one Homesteader made over $5000 from this enterprise.

HOMESTEAD'S SERVICE CENTER

The crossroad which is defined as "the town" is exclusively a service center for the forty-eight farms. Nothing is produced or manufactured there; each place of business exists to provide goods and services for the outlying bean farms. In 1950 there were fourteen service institutions providing total economic support for ten families and partial support for eight families. These institutions may be listed as follows.

1. Combination bean warehouse and filling station.
2. Repair shop specializing in farm machinery.
3. Repair shop specializing in automobile repairs.
4. Filling station.
5. General store.
6. Combination general store and *café*.
7. Combination drug store and *café*.
8. Farmer's bar.
9. Post office.
10. School.
11. Community church.
12. Baptist church.
13. Community well.
14. Road grader for the state highway department.

The bean warehouse was constructed in 1940 and the operator now manages a filling station in connection with his service of re-cleaning and storing beans. Gasoline is hauled by the operator from a refinery in the eastern part of the state and sold in the filling station for less than the prevailing price in the city. The income from this bean-handling and filling station business varies enormously depending upon the bean crop. In 1946 the warehouse handled 13,000 sacks of beans and the operator probably had a net income of almost $8,000; in the drought year, 1951, the warehouse handled only twenty-five sacks. But the average since 1940 has been 7238 sacks per year. At any rate, the business has provided total support for one family and partial support for another — the wife of the second family is employed as the filling-station attendant.

Homestead has two repair shops. One was an early blacksmith shop which sharpened plow shares and shod horses, and gradually evolved into a shop with an electric welder and other machines for the repair of farm machinery. The other shop was built in 1946 and specializes more in automobile repair. In good years both shops are constantly busy. The first shop provides partial economic support for one family which also operates a farm; the other is the sole economic pursuit of one of Homestead's "bachelors." The automobile repair shop had a net income of "about $1200," the other shop "about $1160" in 1949.

The second filling station in Homestead has had a record of opening and closing periodically. It has been impossible for the operators to compete with the gasoline prices of the other filling station and it has never been a "really paying proposition." It closed for the last

ECONOMY OF HOMESTEAD 53

time in the spring of 1951 when the operator ran off to Kansas City with another man's wife.

The large general store has done a reasonably good business in recent years. It was originally built in 1937 and operated by the owner of a large mercantile house in one of the nearby towns. In 1942 the store was purchased by one of the local Homesteaders who operated it until he moved to the Rio Grande Valley in 1948 and rented the store to the present operator. This store carries groceries, clothing, notions, and hardware. The present operator reports that he sold about $16,000 worth of merchandise in 1950. His rent is $540 a year and his retail prices are about 20 per cent above wholesale costs, making his net income approximately $2500 in 1950. The family lives in the back of the store and derives its total income from this business.

In 1948 a second store combined with a small *café* was opened for business. Previously this house of business had been a *café*. In 1950 the business was operated by an elderly Homesteader and his family who lived in the back of the store. An estimate of its income cannot be ventured, but it is known that sales were substantially less than those of the general store. In the drought year of 1951 this family sold the business to the operator of the general store and moved away from Homestead.

The combination drug store and *café* was doing a substantial business in meals, drugs, magazines, and soda-fountain products in the early months of 1950, when a highway crew was working on the road near Homestead. The business dropped off markedly by late 1950, and in 1951 the store closed and the family operating it returned full-time to their ranching and farming activities. At the peak of its business, this store provided only partial income for one family.

Although there had been a small bar in Homestead during the 1930's, the present bar was not opened for business until 1946, when a Spanish-American entered the liquor business in the community. The bar is located at the edge of the town and is lodged in a one-room frame building. Liquor sales consist chiefly of beer and wine during the week and bottles of whiskey for dances on Saturday nights. The bartender reports a net income of $3000 in 1950, and his business is a full-time occupation.

Homestead's post office,[6] established in 1936, is now run by the wife of one of the farm operators and provides partial economic support for this family. The postmistress' salary depends upon her sales of stamps and in 1950 was $1815.

The school system employs five teachers, six school-bus drivers, a janitor, and a cook for the hot-lunch program, and had ninety-four students enrolled in 1950. Three of the teachers were wives of local farmers, and one of the men teachers also operates a farm in Homestead. The other man teacher, who was also the principal and received a higher salary, was the only teacher whose salary from the school was the means of full economic support for a family. In 1949–1950 these salaries averaged $3200 per annum. The janitor was paid $85 a month, and the cook, $80. The janitor was also employed as the community well tender and the cook was the wife of a local farmer, so that these salaries were only partial family incomes. The school-bus drivers were also local farmers, so that their average monthly salary of $175 was supplementary income. One of the schoolteachers was also employed as the teacher of the Veterans Administration school in farm management at a salary of an additional $200 a month. There were ten veterans in this training program, drawing subsistence checks of $65 to $90 a month.

The Community church is visited by a Presbyterian missionary twice each month, who receives a salary of $1900 per annum and is furnished a house and automobile. In 1950 the Baptist church had a resident "preacher" who received $135 a month for his ministerial services. One hundred dollars was provided by the Baptist missionary organization; the other $35 came from monthly contributions of the Homestead Baptists.

The community well is managed by the Homestead Coöperative Association, which was organized in 1940 and now has fifteen members. Shares in the coöperative cost $5 each and a person must have thirteen shares to qualify for full membership. The management is in the hands of a three-man board of directors who in turn hired a well manager. Water is sold at the rate of $.10 a barrel or at the flat rate of $2 a month per family. The well manager received one-half the amount of total sales and had been earning $15 to $30 a month for starting the pump, collecting the money, etc. In 1950 the coöperative was in the red because "some people were hauling water without signing up" and "others would not pay their bills." It was decided to install a coin-operated meter which now measures out a barrel of

ECONOMY OF HOMESTEAD 55

water when a dime is inserted into the coin slot. At the same time an automatic switch was installed so that the services of a well manager were no longer needed.

Finally, the state highway department now maintains a road grader in Homestead and employs one of the local Homesteaders for $175 a month to grade the roads in the vicinity.

In addition to the formal service institutions in Homestead center, there are a large number of part-time occupational pursuits which add to the income of particular families. These specialized occupations may be listed as follows.

1. Two men who drill water wells in Homestead and vicinity for $3 a foot.
2. One water witch who "locates" wells for the farmers in the community and receives from $5 to $10 for his services.
3. One plumber who installs running water and bathroom fixtures in Homestead houses.
4. One barber (the bartender) who cuts hair at his bar.
5. Two farm wives who also function as hair dressers.
6. One Notary Public who charges $.25 for notarizing documents.
7. Five musicians who play violins and guitars for the local dances and receive the admission money of $1 from each man who dances (less $1 which is paid to the janitor for cleaning the schoolhouse after the dance).
8. One combination nurse and midwife who ministers to the sick and delivers babies.
9. Four men who work by the hour as carpenters.
10. Three men who work by the hour as plasterers.

HOMESTEAD'S MARKET TOWNS

Beyond the Homestead service center are the various market towns which form the principal links with the outside world. The most important of these towns is Gallup, seventy miles north of Homestead. Founded in 1882 as a division point on the Santa Fe Railroad, Gallup has become the principal market and trading center for a large section of northwestern New Mexico and now has a population of approximately 10,000. The ethnic composition is diverse; the town contains groups of South Europeans (principally Slavs and Italians), Spanish-Americans and Mexicans from Old Mexico, Navahos and Pueblos who have moved into town, and "Old American" families from the Middle West.[7]

Across the state line in Arizona, at a distance of sixty miles from

Homestead, is St. Johns, a half Mormon and half Spanish-American community of approximately 1500. Although some grocery shopping is done in this community, it is more important as the location of the John Deere and Ford tractor dealers who sell most Homestead farmers their tractors and other farming equipment. The tractors are trucked as far as the foot of the "rim" south of Homestead. The farmer then meets the truck and drives his new tractor up the steep road into Homestead.

South of Homestead forty-five miles is the town of Quemado which is a Texan and Spanish-American ranching town of approximately 400 people. Quemado is a secondary shopping center for the Homesteaders.

A decade ago the largest store in Homestead was owned by a merchant in Grants — a town of 2500 located one hundred miles across the mountains northeast of Homestead. At this time part of the bean crop was sold in Grants and substantial shopping was done there. A few Homesteaders still carry on some business with Grants' tradespeople.

Although Gallup is the principal urban center for shopping, banking, marketing, and recreation, the Homesteaders must go 177 miles east to Los Lunas — their county seat — for farm and county business. They feel that they have never had sufficient influence in county affairs, being Democrats in a county controlled politically by Republican Spanish-Americans, and most of them are strongly in favor of re-drawing county lines to place Homestead in the same county with Gallup.

The "big city" is Albuquerque, located 175 miles east of Homestead, with a population of over 100,000. Nearly every family makes a few trips each year to the city, often in connection with a business trip to Los Lunas

PRICES AND INCOMES

The preceding detailed facts and figures on the economy of Homestead may be summarized and focused upon the critical problems by an analysis of the price and income structure of the community. Since many rural businessmen and farmers keep no record, or no accurate record, of their transactions, these summary figures are necessarily estimates, based upon what facts are known and are available to the observer. For example, although a farmer may not

ECONOMY OF HOMESTEAD

be able to tell you what his income was in a particular year, he can tell you how many sacks of beans he raised and sold, how much he obtained for the sale of a given piece of land or livestock, etc.

The basic resource in this farming community is the land itself, which in recent years has risen sharply in price. In the early 1930's land could be purchased for $1.00 an acre, and in 1940 Goodsell reports that unimproved land was valued at from $1.00 to $1.50 an acre.[8] By the middle 1930's improved land was selling for $2.00 to $3.00 an acre. However, the sharpest rise in land prices came with the marked increase in bean prices during and immediately after World War II. In 1950 the prevailing price for improved land was said to be $10.00 an acre, and accurate figures on the sales of the most recent five sections to be sold in Homestead show an average price of $9.72 an acre.

The present tax structure does not reflect these sharp rises in land prices, for at the present time grazing land is assessed at $1 an acre, farming land at $5 an acre, and the taxes are about $18 to $20 a section for most of Homestead's farms, which contain varying proportions of grazing and farming land.

Because of chronic environmental threats from drought and killing frosts, income fluctuates enormously from year to year with the fluctuations in bean production. The total acreage under cultivation in Homestead has increased from an estimated 6000 acres in 1940[9] to 11,213 in 1950, but total bean production has not shown the same increase. Instead, bean production has shown wide fluctuations over the past decade, which reflect the natural conditions of precipitation and the price of beans on the market (see Table 1). It can be readily seen that over the past nineteen years there have been seven farming years (1933 to 1937 inclusive, 1946, and 1948) in which bean production was well above average and bean prices were average to exceptionally high. On the other hand, there have been at least five years (1938, 1939, 1944, 1945, and 1950) in which there were crop failures or near failures and bean prices did not compensate for the drop in production. The other seven years have been average years when farmers "just got along." If the analysis were extended to include 1951, one more extremely bad year would be added to the list, for in 1951 only twenty-five sacks of beans were produced. The totals would then stand as seven good years, six bad years, and seven average years over the full two decades that the community has been in existence. The summary statement made by Goodsell in 1940 for the plateau area as a whole to the effect that "shortages of native forage as well as cultivated crops occur about 3 years in 10 because of drought conditions" [10] would hold accurately for the total period since 1931.

For a somewhat more detailed estimate of the source and amount of Homestead's annual income, I have selected the year 1949, for which I was able to obtain reasonably accurate information. No strong claim is made for the typicality of this year in reflecting the average conditions in Homestead. However, it is clear that 1949 was neither one of the seven good years nor one of the six bad years. Eighty-three hundred sacks of beans were handled by the local warehouse, a number close to the average of 7238 passing through this warehouse over the ten-year period of its operation. The yield per acre was approximately 2.5 sacks, which is less than the average yield of 3.5 sacks in the period from 1932 to 1940.[11] It is not known definitely whether this lesser yield is due mainly to precipitation conditions or to a decrease in the fertility of the soil for bean production over the past decade, but the fact that rainfall was only slightly less than average for the decade and that given fields are known to produce less now than they did previously suggests that there has been some decrease in fertility. At any rate, 1949 is to be regarded as an approximation to an "average year." The estimated income figures for Homestead in 1949 can be summarized as follows: [12]

	Gross	*Net*
Total income	$209,900	$157,200
Average family income (all 61 families)	3,441	2,577
Income of 13 "town" families	31,400	27,700
Average "town" family income	2,415	2,131
Income of 48 farm families	178,500	129,500
Average farm-family income	3,719	2,698

The sources of income in 1949 were as follows:

Farming activities	$119,900 (57%)
Sale of beans	79,800 (38% of total income; 66% of farming income)
Other, including livestock sales	40,100 (19% of total income; 34% of farming income)

ECONOMY OF HOMESTEAD

Service activities	65,000 (31%)
Wagework	9,500 (4.5%)
Government payments	15,500 (7.5%)
TOTAL	209,900

Farming activities in these figures include income from both crops and livestock; wagework includes all compensation from work done by the hour or day in or away from the community; government payments include subsistence checks for Veterans Administration training programs, PMA payments for conservation of fields, and disability payments to veterans. It is instructive to note that although the sale of pinto beans provides the largest single source of total income in Homestead (38 per cent), the income from service activities is almost as important.

Although the average net family income in 1949 was $2577, there were important differentials in income among the families. The range was from $8000 to less than $600. The following tabulation provides a more precise picture.

Number of families	Net income for 1949
7	Over $5000
5	$4000 to 5000
5	3000 to 4000
15	2000 to 3000
25	1000 to 2000
4	Less than 1000

If 1949 is considered an approximately average year and these net income figures are compared with the average incomes earned in wagework away from the community during 1950–1951, a number of highly interesting facts emerge. As indicated before, the average earning for wagework during this period was $80 a week, or an estimated $3900 per annum (correcting for time lost due to changing jobs, inclement weather, and illness). The lowest wage received by any family was $75 a week, or an estimated $3600 per annum.

The higher cost of living in the wagework situation in the nearby cities or in agricultural employment in Arizona or California is the result of two basic factors that are not present in Homestead: rent and the necessity of purchasing one's entire stock of food. The average rentals paid by Homestead families in the wagework situation in 1950–1951 were approximately $45 a month, and the cost of food in urban centers approximated $40 a month in excess of what is paid for food while living in Homestead. In other words, the average wageworker away from the community earned about $2880 a year in real income, or approximately $300 more than it would be possible for him to earn in Homestead in an average year. Indeed, only some eighteen of Homestead's families are in a position to exceed this income. According to our previous calculations of the farming risks in Homestead, a family would have only one chance every three years to better this income while engaged in the farming of pinto beans. Furthermore, in three years out of ten a family's income would be markedly lower in Homestead than from doing wagework elsewhere, assuming present economic conditions.

Despite these compelling economic facts, every family which moved away from Homestead in the winter of 1950–1951 returned to plant a crop and "try it again" in the spring of 1951. The implications of these facts will be treated in a later chapter, but it may be said here that these data appear to provide solid evidence for the operation of value-orientations over and above sheer economic considerations.

PART II THE VALUES AND THE COMMUNITY

3

Hopeful Mastery Over Nature

The first wave of western pioneers, composed of explorers, trappers, and mountaineers, was characterized by the marked extent to which these men blended their activities into the environment and made adjustments to existing natural processes. Later agricultural settlers were characterized by a drive to dominate the natural environment and to bring it under greater human control. The settlers of Homestead were the type of frontiersmen who defined nature as something to be mastered, controlled, and exploited by man for his own ends and material comfort.

A few of the Homesteaders, under the influence of soil conservation programs, have come to hold an attitude of greater respect for the land and its resources — an orientation which approximates the man-in-tune-with-nature attitude of the Pueblo Indians. In the face of continued droughts, sandstorms, and killing frosts which damage the land and destroy the bean crop, other Homesteaders manifest an orientation of hopelessness of ever doing anything to control these natural conditions and feel, like the Spanish-Americans, that it is best to take things as they come. But the predominant attitude is one of hope in man's rational-technological ability to eventually overcome the vicissitudes of nature — tempered by a realistic recognition that (at least for the present) farming in this arid land is something of a gamble, and that one has to be lucky as well as skillful in order to "make a crop."

To say that an orientation of hopeful mastery over nature is stressed by the Homesteaders does not mean that all the techniques used are rational-technological from the point of view of modern science. The use of machinery in farming operations, of birth-control techniques and modern medicines, are examples of rational methods. But, in addition, there are "magical" or "ritual" techniques, such as the folk belief and practice of "water witching," "planting by the moon," reading the "signs" of the zodiac, etc. Nevertheless, the central emphasis is still upon "controlling" natural processes, whether the means are "rational-technological" or "magical" from the observer's viewpoint.

In order to illuminate the problems which are currently confronting the Homesteaders in their relationships to nature, a sequence of historical situations which has an important bearing upon the present value-orientations of the community will first be described.

THE WINTER OF THE BIG SNOW

On November 21, 1931, the deepest snowfall on record (thirty inches) blanketed the Pueblo Plateau country. Old-timers expected the snow to melt away rapidly as had other snows which fell so early in the winter season, but the weather turned bitterly cold, and three days later there was a second heavy storm. The snow reached a depth of almost five feet and remained on the ground all winter. At the time, Homestead had barely been settled, and only a few families had farmed crops in small clearings in the woods. Other Homesteaders had just arrived and were still in the process of building cabins and getting started in the new land. The community was almost completely isolated from the outside world for long periods of time, and the Homesteaders subsisted on diets of gravy and homemade bread, supplemented by jack rabbits, porcupines, and pinyon nuts. Although there were no human deaths in Homestead, the toll in livestock was heavy, and some Navahos in the area suffered starvation and death by freezing.

The "winter of the big snow" is now remembered as the earliest event of importance in the history of the community and has become a symbol of the Homesteaders' success in combatting the natural environment. The event was formalized in the traditions of Homestead when on November 21, 1941, "The Loyal Order of Homesteaders" celebrated, with a "10th Anniversary Dinner Party," the date of the "big snow." Speeches were made commemorating the heroic pioneering days during that early fateful winter.

"NEW DEAL" GOVERNMENT PROGRAMS

Following the "winter of the big snow" there were several critical years during which the Homesteaders faced problems of low farm prices on the one hand and the chronic risks of crop failure due to drought and frost on the other. But with the initiation of the "New

HOPEFUL MASTERY OVER NATURE 65

Deal," many government agencies began to assist the Homesteaders. While some of these government programs were approved of, others aroused strong opposition in the community.

In 1933, a program for building and repairing roads was initiated in the county by the Civil Works Administration, which provided employment for many of the people of Homestead. By 1936, others were employed by the Agricultural Adjustment Administration for the construction of stock water tanks, terraces, and water spreaders; by the Soil Conservation Service for the construction of dams, terraces, and water spreaders in the area. Also initiated during this time were relief payments to indigent families by the New Mexico Department of Public Welfare, and grants to needy farmers by the Farm Security Administration.

There are few figures available on the amounts paid directly to families in Homestead, but some indication of the local situation may be drawn from Goodsell's report on the Pueblo Plateau as a whole, including Homestead. Up to 1940, federal and state agencies had disbursed approximately $94,000 in grants, wages, relief payments, and agricultural program payments in this area.[1] Forty-two per cent of the total was expended by the Works Progress Administration and the New Mexico Department of Public Welfare. Most of the funds from these two agencies went to nonfarm people; however, many Homesteaders worked in the WPA program, and in 1936, twenty-nine of the Homestead families, comprising about one-third of the population, were on relief. These families on relief were almost all from the group of Homesteaders who, when they settled in the area, arrived without resources.

Of the three agricultural agencies, only the Farm Security Administration distributed a large portion of its funds, 82 per cent, to the farmers surveyed by Goodsell in the Pueblo Plateau.[2] The expenditures of the Agricultural Adjustment Administration and the Soil Conservation Service went primarily to ranchers and wageworkers, respectively. Consequently, the farmers surveyed received only 27 per cent of the total amount disbursed by these three agencies. The payments averaged $142 per farm family, or $34 per capita.

The combined programs of the five government agencies offered substantial assistance to the community in these critical years; the total expenditure for all the Pueblo Plateau communities for this period was $53 per capita. However, large public expenditures were also generously provided for other agricultural areas during the depression, and the figures for other areas of the Southwest will show the amount for Pueblo Plateau to be minimal by comparison. In eighty-five counties in the Southern Great Plains, public expenditures by similar agencies during the same period amounted to $225 per capita; in Curry County, New Mexico, they were $217 per capita; and in Harding County, they were $445.[3] In other words, although Homestead may have been "saved" [4] during the depression by these public expenditures, so were countless other, more stable agricultural communities on the Plains and in the Southwest.

Although the financial assistance outlined above was accepted and appreciated, there were other government programs initiated in the 1930's to which the Homesteaders were violently opposed. The Taylor Grazing Act, passed in 1934, effectively stopped further homesteading in the area and permitted ranchers to graze cattle on the public domain, thus defeating hopes for rapid expansion of small farms in Homestead (see Chapter 4).

Later in 1934 the Farm Security Administration, working in conjunction with the Land Use Division of the Department of Agriculture, developed a plan for resettlement of the Homesteaders. They hoped to purchase the land in the Homestead area, turn it back to stock-grazing use, and allow the Homesteaders to use the purchase money to buy farms in a newly irrigated farming area in the Middle Rio Grande Valley. Almost without exception, the Homesteaders resented the conclusion that they had settled on "submarginal" land, and they refused to be resettled. The objections were expressed forcefully, according to one of the original settlers who reported:

The people said, "They'll have to take a shotgun to move us out of here. We're going to stay here just as long as we damn please."

Several years later a number of the Homestead farmers moved to the Rio Grande Valley and purchased farms "and did the very same thing the government wanted them to do in 1934." But they did it "on their own" and were not "forced to do it by those damned government agents."

In 1936 the same government agencies proposed a "unit reorganization plan" which would have enabled the Homesteaders to acquire additional tracts of land on which to run more livestock, and hence make them less dependent upon dry-farming. The plan called for the use of government funds to purchase large ranches near Homestead which would be managed coöperatively by a board of directors selected by the community. The scheme collapsed while it was still in the planning stages, because it became evident that each family in Homestead expected to acquire its own private holding on the range and not participate in a coöperative arrangement.[5]

Nevertheless, government agencies continued to be concerned for the future welfare of the Homesteaders and for the conservation of the land they occupied. The Interdepartmental Rio Grande Board [6] recommended that no new agricultural entries be allowed in the Pueblo Plateau area and that a study be undertaken which would assist the interested agencies in formulating policy. The resulting coöperative investigation included a soil survey, a study of ground water resources, and an economic appraisal of dry-farming.[7] The general conclusions were, in part:

> The [Pueblo] Plateau area requires an economy built primarily on grazing, because so much of the land is either not physically cultivable or would not return sufficient income from crops to justify tillage. As only small scattered tracts of the lands are arable, crop farming can never become the predominant industry. Even the cultivable acres do not afford first-rate opportunities for straight crop farming because of the sparsity and wide variation in annual and daily distribution of rainfall.[8]

The reports further indicate that if people persist in farming they should have larger acreages (some reports say at least four sections per family) to make it possible to combine farming with livestock grazing. Various combinations of farming and livestock raising were also recommended for different types of land in order to maximize income. But these final reports were not seen by the Homesteaders until it had become mainly an academic question,[9] since the possibilities for individual small farmers to acquire additional land were effectively cut off by the emergence of large ranch holdings on all sides of the community.

Meanwhile, the Homesteaders were also receiving substantial aid by establishing credit relationships with banks in the nearby towns. In particular, a prominent banker in Gallup, familiar with the problems of the Homesteaders, came to have the greatest admiration for the "sincere and enterprising nature" of the people. The banker, who had himself risen to prominence from a humble Southern background, sympathized with their difficulties and made a genuine effort to help these Texas and Oklahoma bean farmers who tried to better themselves by "their own initiative and own efforts." The extension of loans through years of crop failures and low farm prices by this banker has been important in the survival of Homestead. But through the critical days of "battle" with the government, which had defined their community land as "submarginal" and unsuitable for agriculture, there emerged in the Homesteaders a sense of mission in life: to demonstrate to the experts in the Departments of Agriculture and the Interior that the Homestead area *is* farming country and that they can "make a go of it" in this semiarid land. They point to the fact that Pueblo Indians made a living by farming in the area long before the white men arrived. There is a general feeling that somehow the surveys and investigations made by the experts must be wrong. They insist that the Weather Bureau has falsified the rainfall figures that were submitted by the Homestead weather station in the 1930's, and indeed they stopped maintaining a weather station because they felt that "the figures were being used against us."

Despite the vicissitudes of the weather and the efforts of the government to resettle the Homesteaders, the community moved forward with confidence that the local environmental situation could eventually be mastered and the agricultural system gradually expanded in acreage and production. The farms were originally small clearings of a few acres in the pinyon-juniper woodlands, but by 1940 the amount of land under cultivation had reached 6,000 acres and in 1950 was 11,213 acres.

RATIONAL-EMPIRICAL METHODS

The control of nature in Homestead today is most immediately apparent in the use of the machine for farming operations. For the first few years after settlement, the clearing of the land was done by hand, and farming was done with horse-drawn implements. Even at this time, it was evident that the Homesteaders were changing the landscape more than it had ever been changed in the history of human settlement in the area. The first tractor was brought to the community in 1935, and by the early 1940's hand labor and the use of animal power had been almost entirely replaced by the use of

HOPEFUL MASTERY OVER NATURE 69

power machinery in most of the important farming operations. Although Homestead represented the *most recent* cultural development in the area, it was the *first* of the five cultural groups (the others being the Mormons, Pueblos, Navahos, and Spanish-Americans) to become highly mechanized. Today there is more power machinery in Homestead than in any of the nearby communities.

The Homesteaders' strong and rapid advance toward mechanization clearly reflects their definition of the natural environment as something to be subdued by man, and there is relatively great enthusiasm for the new and more powerful machines which are the means for effecting the task. Bulldozers are now used to clear away the pinyon and juniper trees that cover potential farming land, and they are used in the building of tanks and spreader dams for the storage and conservation of water. Wells are drilled with gasoline-powered rigs. The farm tractor, however, is the principal power machine for farming operations, and a Homesteader on his tractor is the most typical expression of the empirical methods of controlling nature in present-day Homestead. Many complex farm implements are used with the tractor: lister-planters, knife-sleds,[10] cultivators, breaking plows, discs, harrows, chiselers, bean-pilers, and combines.[11] Post hole diggers attached to tractors are also utilized for fence building.

The basic operations for farming pinto beans include clearing the land of timber, breaking the sod, listing, planting, knifing, cultivating, piling, and threshing the beans.[12] The sod is broken up with various types of plows, and then plowed with a lister, a process which throws the soil up into deep rows. Lister rows are always cross-wind (i.e., they run north and south in this area of prevailing westerly winds) and are designed essentially to reduce wind erosion.[13]

Another method for controlling the removal of topsoil by windstorms is the common practice of strip cropping. This consists of interspersing either four or eight rows of corn with twelve or twenty-four rows of beans. Both the listing and the strip-cropping practices are encouraged by the Department of Agriculture through its office of the Production and Marketing Administration, located at the county seat. The farmer is paid 30c an acre for listing and 60c an acre for proper strip cropping. On a few farms the slope of the fields is steep enough to warrant contour cultivation, but most of Homestead's farms are level enough to make this practice unnecessary. Furthermore, contour farming increases the danger of wind erosion because some of the rows will inevitably run in the direction of the prevailing westerly winds which (rather than water) cause the basic erosion problems in the Homestead area. Consequently, most Homestead farmers choose not to practice contour cultivation and look askance at the elaborate terraces constructed by the Soil Conservation Service in the 1930's.[14]

After the planting operation, there are usually three cultivations; the first is performed with the knife-sled, and the second and third with a duck-foot cultivator. A few farmers disc their fields before listing them; some use harrows rather than knife-sleds for the first cultivation; four farmers have tried chiseling the land — a process which breaks up the soil to the depth of a foot and opens the ground to receive additional moisture — before listing. Most farmers, however, perform only the basic operations discussed above. At harvest time the crop is cut with knife blades attached to a cultivator or lister beam. The beans are then piled in small shocks for drying, after which they are threshed with a small-grain combine (only three farmers still use a threshing machine rather than a combine).

Obviously this battery of machines used in current bean-farming operations has reduced the need for additional farm labor to a minimum in Homestead. In earlier years when beans were harvested with a threshing machine, coöperative working crews of twelve men were needed for harvesting operations. Only at two points in the agricultural cycle is there still a need for additional labor; (a) if a farmer does not use a bean-piler, the beans must be piled by hand; (b) if there is promise of a good crop and the weed growth is heavy, it pays the farmer to hoe out the weeds near the bean rows which are not cut by the knife-sleds or the cultivators. This additional labor is provided in two ways: exchange labor with one's neighbors or kinsmen or both; or the hiring of additional farm laborers (usually Navahos, but sometimes Spanish-Americans or other Homesteaders) by the day.

Along with the application of high-powered machines to farming operations, there has developed in the past two decades an impressive body of technical knowledge of farming in a semiarid environment. Knowing how and when to plant, how to cultivate to conserve moisture and kill weeds in the same operation, and how and when to

HOPEFUL MASTERY OVER NATURE 71

harvest the bean crop, are all problems that the newcomer to Homestead cannot master in less than three or four years.

With the arrival of the Rural Electrification Administration power line in Homestead in 1949, there has also been mechanization of the farm home. All Homesteaders within reach of the REA have had their homes wired for electricity (in marked contrast to the Spanish-Americans in the nearby village of Tapala, where only one family has installed electricity). This was immediately followed by a wholesale rush to purchase such home appliances as deep freezes, refrigerators, irons, washing machines, coffee percolators, electric water pumps, etc., even though these purchases almost invariably put the Homesteader further in debt, and it was not always clear that the new equipment was really needed for comfortable living.

The enthusiasm for machines also extends to transportation and communication. No Homestead family is now without at least one automobile, and many families have both a pickup truck and a passenger car.[15] Transportation between farms, into Homestead center, and into the nearby larger towns is entirely by automobile. Horses are still kept by a few families, but (except for three farmers who still farm with horses) are used only for rodeo activities and for riding by the youngsters. Every family has a radio, and Homestead looks forward eagerly to having television. Homestead also impatiently awaits the building of a telephone line into the community; meanwhile, one family has installed a "barbed-wire" telephone line which utilizes the wire on farm fences to carry messages between the homes of three related families.

The technical knowledge for the repair and maintenance of most of Homestead's machinery is found within the community. Radios are usually taken to Gallup for repairs, but tractors (and associated implements), automobiles, and home appliances are most often repaired by the farmer himself or by the repairmen in the local shops. The level of technical competence in the operation and repair of complex machinery is markedly high for a small rural community.

THE PROBLEM OF OVER-MECHANIZATION

The interesting question now arises as to whether or not the Homesteaders possess more machines than they need for the amount of land they farm, the amount of traveling they do, the amount of food they have to keep in deep freezes, etc. A close examination of the relationship between the number and size of the machines and the amount of work to be done will show that the Homesteaders are "over-mechanized"; it is also clear that this "over-mechanization" is related to important value-orientations, especially "mastery over nature."

Let us examine the problem by reviewing the Homesteaders' investment in farm machinery. As of 1950, only three of the forty-eight farm operators were using horses; the others all possessed tractors (and the associated implements), and three farmers had two tractors. Approximately one-third of the tractors handled four-row equipment; the others handled two-row equipment. The costs of the tractors averaged well over $2000, and with the associated implements, the total value of each farmer's machinery was at least $4000, as previously noted. Yet even during farming season, these machines were standing idle in the farm yards more days than they were being used in the fields. There was general agreement among the Homesteaders interviewed that, with careful scheduling, it would be feasible to farm at least two of the larger farms with a single four-row tractor, and two of the smaller farms with a single two-row tractor. But there was also general agreement that "everyone should have his own tractor." The reasons for this position were also clear: with his drive to "control and subdue" nature, each farmer felt that he needed a tractor (the larger and more powerful, the better) to exercise this control. As one Homesteader expressed it:

> I like a big four-row tractor under me when I'm out there in the fields. It's got the power you need. I don't see why anybody wants to fool around with one of those little Ford tractors.

In addition, the Homesteaders realize that with their stress on individualism the coöperative use of the farm tools on two or more farms would not work. "Coöperative farming wouldn't work because everybody's got a different idea about how to do the farming," asserted one of the old-timers. Finally, it is plain that the possession of new and more expensive machines has become an important symbol of individual prestige and achievement in the competitive social order. The whole conception of a "successful" farmer involves the possession of these machines, and those farmers with four-row John Deere tractors and full sets of implements are "the big operators." Those farmers with smaller Ford tractors and less expensive implements are "just scratching around."

HOPEFUL MASTERY OVER NATURE 73

A similar situation prevails in regard to the possession and use of automobiles. Whereas the nearby Navahos and Pueblos travel to Gallup in cars loaded with relatives and neighbors, the Homesteaders now travel the seventy miles to Gallup (and elsewhere) in nuclear family groups. On any one day there will be several half-filled cars leaving Homestead for the city at approximately the same time of day and returning in the late afternoon at almost the same time. Since the community is small, almost everyone knows who else is going into town on a given day. Yet, as is true with Americans generally, the possession and individual use of an automobile is regarded as a necessary aspect of the way of life, regardless of the cost in terms of gasoline and depreciation.

A comparable phenomenon was observed in the purchase of $400 to $500 deep freezes when the power line was built in 1949. Many large sixteen-foot units were purchased on time payments and have since been completely empty many months of the year, and only partially filled the rest of the time. It is clear that two or more families could have shared a single deep freeze, but because of the value-orientations of Homestead this was simply not feasible.

The technological development in Homestead is disproportionately elaborate, viewed in the context of the objective work and economic needs of the community. Over and above sheer economic considerations, "the machine" acquires a special significance in the Homesteaders' value-orientations, with a resultant "over-mechanization": machines are used not only to extend control over the natural environment, but also as crucial symbols of prestige and individual achievement in the community.

WEATHER CONTROL

Although power machinery and technical knowledge give the Homesteader confidence, he is still faced with the critical problems of drought, frost, and wind erosion. Strip cropping and running the crop rows cross-wind help to reduce wind erosion; the use of early maturing bean seed reduces somewhat the danger of frost; and the planting of the more drought-resistant types of crops helps to reduce the risks in dry years. But there is obviously a point beyond which the Homesteader exercises no control over these problems with his present technical knowledge. Nevertheless, there has been much talk about the use of smudge pots (like those used in the orchards of California) to protect the bean crop from early autumn frosts. The idea has proved to be impractical because of the large number of smudge pots that would be required to cover the fields.

There has also been widespread discussion among the Homesteaders of the possibilities of "rain making" by the use of dry ice or silver iodide. During the drought years of 1950 and 1951 a rain-making company had contracts to seed clouds over much of New Mexico. The closest mobile generator was located at Pueblo, and two Homesteaders (with large landholdings in the community) joined in the rain-making program, paying $10 a section for the service. Like the physicists and meteorologists who are studying the rain-making experiments, the Homesteaders were divided (although for different reasons) in their opinions of this technique. A minority felt that it was "interfering with the work of the Lord" and that "God is the only one who has the power to make it rain." But the majority thought that artificial rain making would eventually become effective, and some went so far as to say that "in ten years you will be able to call up and tell them when you want rain and how much and they'll provide it for you." When asked whether seeding the clouds would be "interfering" with the Lord's work, the leading proponent of rain making replied:

No. The Lord will look down and say, "There's those poor ignorant men. I gave them the clouds, the airplanes, and the dry ice and they don't have the sense to put them together!"

Along with the discussion of rain making, there has also been serious consideration of the possibilities of obtaining irrigation wells in Homestead and thereby relieving the drought problem. In 1951 two such wells were attempted, but in neither case were sufficient quantities of water obtained to make pumping and irrigation economically feasible. Since proper geological conditions do not exist for the development of artesian wells and since there are no feasible locations for irrigation reservoirs (and no mountain snow packs to feed reservoirs), it seems unlikely that the Homesteaders will ever be able to shift to irrigation farming.

MEDICAL PRACTICES

The health of the Homesteaders is relatively good. During the winter almost everyone has a "cold" once or twice, and there are occasional

HOPEFUL MASTERY OVER NATURE

cases of pneumonia. In the summer there are often various types of dysentery (of short duration) among the population. The "children's diseases" of measles, whooping cough, chicken pox, and mumps have occurred in the community during the past few years. There has been one death from polio and one death from a heart attack, but there have been no reported cases of cancer, tuberculosis, or other diseases that are widespread elsewhere. Despite the distance from the hospital (seventy miles over rough roads), the Homesteaders do not worry much about their health. Certainly, as compared to the anxiety about rainfall and crops, the concern about illness is negligible.

In the treatment of illness there is some use of home remedies and of patent medicines purchased in the local stores, but by and large the Homesteaders are oriented strongly toward the use of the latest medicines and medical facilities when sickness does occur. Some illnesses are treated by the elderly "Doc" who used to be a practicing physician in Texas and is now the mail-carrier between Quemado and Homestead. In other cases the patient is immediately transported to one of the physicians in town. Innoculations against children's diseases are administered in the school by the county nurse, and most of the younger parents take their preschool children to the large town some distance away for the recommended series of innoculations.

In the early days of settlement, childbirth occurred in the homes with the assistance of a midwife, and there is still one practicing midwife in the community. But within the past decade all babies have been delivered in the hospital in the city, and the midwife ministers mainly to the Spanish-American women in Tapala. Modern methods of birth control are also practiced by almost all of the married couples in Homestead.

RITUAL METHODS

Malinowski has insisted that every human culture possesses both sound scientific knowledge for coping with the natural environment and a set of magical practices for coping with problems that are beyond rational-empirical control.[16] Despite the tendency of most rural agricultural experts and educators to assume that such practices are largely matters of history, or at least persist only as "superstitions" among the most unenlightened farmers, the rural American scene is still characterized by a fair share of magic associated with farming operations. In *Plainville, U.S.A.*, James West found that "many magical practices still exist for planting crops, castrating livestock, weaning, gardening, girdling trees." [17] Carl Taylor reports that he has gathered 467 different signs and superstitions which are known and to some extent believed in rural communities.[18] In Homestead, too, these magical practices exist side by side with the rational-empirical methods described in the preceding section.[19] Compared to the rational-empirical methods utilized by the Homesteaders, these ritual practices play a minor role in the culture but they are still widespread and persistent enough to require analysis and explanation.

From the writings of Pareto, Malinowski, and Weber, and the present generation of theorists — notably Parsons, Kluckhohn, and Homans — there has emerged a general body of theory concerning the function of ritual in the situation of human action.[20] Briefly stated, the theory is that when human beings are confronted with situations which are beyond empirical control and are, therefore, anxiety-producing both in terms of emotional involvement and in terms of a sense of cognitive frustration in the situation, they respond by developing and elaborating nonempirical ritual which has the function of relieving emotional anxiety and of making sense of the situation on a cognitive level.[21]

In a recent publication, Kroeber has questioned the universality of this relationship. He has pointed out that the Eskimos, who live in a far more uncertain and anxiety-producing environment than do Malinowski's Trobriand Islanders, have little ritual as compared with the Trobrianders, whereas according to Malinowski's theory, one should expect more ritual among the Eskimos.[22] Kroeber further indicates that the Arctic environment is so severe that had the Eskimos devoted much time and energy to the development of ritual patterns they would have perished long ago. This latter point is sound, but further analysis of Eskimo culture may reveal that although there is little elaboration of ritual (due to the exigencies of the severe environment), the ritual patterns which do exist are still clustered around foci of the greatest uncertainties in Eskimo life.

Other writers, notably Radcliffe-Brown,[23] have raised the issue as to whether rituals create anxiety (when they are not performed, or

The Homestead Bar Is a Popular Loafing Place — by Custom for Men Only

As Population Declines, Former Businesses Close Up

Village Store Is a Place of Business — and Visiting

HOPEFUL MASTERY OVER NATURE 77

performed improperly) rather than alleviate anxiety. Homans has treated this problem in terms of "primary" and "secondary" rituals focused around "primary" and "secondary" types of anxiety. "Primary" anxiety describes the sentiment humans feel when they desire the accomplishment of certain results and do not possess the techniques which make them certain to secure these results. Under these circumstances, humans tend to perform certain actions which have no practical result and which we call ritual. But once ritual procedures alleviating the "primary" anxiety are established as the customary way to do things, a "secondary" anxiety may result when the rites themselves are not properly performed. "Secondary" types of ritual then arise which may be called purifications and expiations.[24]

Kluckhohn has carried the analysis further by demonstrating (in the case of Navaho witchcraft) that ritual patterns have both a "gain" and a "cost" from the point of view of the continued successful functioning of a society.[25] Thus while witchcraft patterns are "functional" from the point of view of providing culturally defined channels for the expression of anxiety and release of aggressive impulses, they also have the "dysfunctional" aspects of adding more things to be feared and of resulting, occasionally, in acts of violence and in punishment of guiltless individuals.

To this development of the conception of ritual I should like to add the theory that ritual patterns which emerge initially as responses to critical areas of uncertainty in a given situation of action are elaborated and reinterpreted according to certain selective value-orientations of the culture.

We can employ, then, a dynamic conception of ritual which includes the following considerations: ritual patterns develop as a response to emotional anxiety and cognitive frustration in a situation of uncertainty; but ritual patterns come to have both "functional" and "dysfunctional" aspects (both a "gain" and a "cost") for the continuing existence of a society as the patterns are elaborated and developed in terms of the selective value-orientations of a given culture.

In Homestead there are three types of folk ritual concerned with farming operations: (*a*) water witching; (*b*) the use of the phases of the moon to predict the weather and determine planting times; (*c*) the use of signs of the zodiac. All three types of ritual are closely related to areas of empirical uncertainty in farming and raising livestock. The water witching technique provides assurance in the uncertain procedure of locating a shallow water well; the observation of the moon, fogs, etc., to predict weather is supposed to aid in determining planting times; the signs of the zodiac guide a great range of tasks such as castrating and dehorning cattle, and weaning calves.

WATER WITCHING

The phenomenon by means of which one is supposed to find underground supplies of water through the use of a divining rod has a respectable antiquity in Western culture, and it is highly probable that the basic ideas derive ultimately from ancient divining practices which are still widespread among the nonliterate cultures of the world.[26] The "rod" is mentioned many times in the Bible, especially in the books of Moses, in connection with miraculous performances. As early as the first half of the sixteenth century, the divining rod in its present form was in use in Germany. From Germany the technique spread to England and from there to the New World and to such regions as Australia and New Zealand.

The core of the water witching pattern as it is now found in Homestead culture may be described as follows:

(*a*) *Equipment*. The most common item of equipment utilized is the witching stick, typically a Y-shaped twig cut from a peach, juniper, or pinyon tree. The two forks vary from fourteen to eighteen inches in length, and the neck from four to eleven inches. The diameter of the stick may vary from an eighth of an inch to almost an inch.

(*b*) *Technique*. The most common technique is to grasp the two branches of the forked twig, one in each hand, with the neck (or bottom of the Y) pointing skyward. Usually the twig is grasped with the palms of the hands up, and the twig is placed under tension in such a way that the slightest contraction of the muscles of the forearm or slight twist of the wrists is sufficient to cause the twig to rotate toward the ground. The water witch takes off across country on foot and when he walks over an underground supply of water, the witching stick is supposed to dip down.

(*c*) *Ideology*. Like most ritual patterns, water witching carries with it an elaborate mythology, the core of which involves two

HOPEFUL MASTERY OVER NATURE

aspects. The first concerns the dowser's* definition of the geological situation and can be summarized by the belief held by dowsers that underground water occurs in two forms: *sheet water* which underlies a total area, and *water veins* which may vary from "the size of a pencil" to "underground rivers" and which run through the earth like the veins in the human body. The most elementary knowledge of ground-water geology is sufficient to prove that this dowsing conception bears little relation to the facts of underground water supplies. The second aspect of the mythology concerns the many and varied "explanations" advanced by dowsers for the efficacy of the technique and the rationalizations which are provided to account for failures. These typically take the form of attributing the failure either to faulty equipment (e.g., "I couldn't find a straight stick that day") or to some aspect of the situation which negates the findings of the dowser (e.g., "I found I had a knife in my pocket which short-circuited the electric current"). Often the failures are forgotten altogether, and the most frequent stories told about dowsing involve the cases in which a farmer tried drilling without dowsing and obtained a dry hole; then he hired the water witch and succeeded in obtaining good water on his farm.

(d) *Institutionalization of Role.* The water witch occupies a special role in the community. In the first place, the basic ability to do witching is believed to be a skill which is innate and can only be "discovered"; it cannot be acquired by training or experience. To be sure, it is necessary for a person to have observed (or to have heard about) the basic techniques, but it is impossible for one water witch to impart the essential ability to another. One is either born with it or is not.[27] By virtue of this innate ability, the dowser assumes a specialized role which is recognized by the total community, both by the proponents and by the adversaries of dowsing. Secondly, the water witch is almost always paid for his services (usually $10 in Homestead), and derives substantial amounts of income from his witching activities.

In the first chapter, I have described the geology and the ground-water resources of Homestead and indicated the critical problems

---

* In current usage, the words "dowsing" and "divining" are more common in the literature and are used by rural people along the Eastern Seaboard; "witching" is the more common term in use by rural folk in the South, Midwest, and Far West.

involved in the development of wells for livestock and household use. When the Homesteaders first arrived, they found it necessary to haul water in barrels by team and wagon from lakes or springs at some distance from their farms. A few tried drilling wells in these early years, but it was soon discovered that while in some places water could be struck at shallow depths (80 to 100 feet), in other localities dry holes were the only result after drilling over 500 feet. At a cost of $1 to $3 a foot for drilling a well, few Homesteaders were willing or able to take the risk of drilling on their farms.

It was in these circumstances that one farmer suddenly "discovered" in 1933 that he had the power to witch for water. He had observed others witching when he was a boy in Texas, so one day he simply cut a forked twig from his wife's peach tree and tried out the technique as he remembered it. He found two water veins on his farm, traced them to a point where they crossed each other, and had a successful well drilled at this spot to a depth of 230 feet. He rapidly achieved community-wide reputation as a water witch and "successfully" witched eighteen wells in the next few years.[28] During the same period, however, he dowsed five locations where dry holes resulted after drilling, and he often missed the depth by as much as 200 to 400 feet. For, in addition to using the common technique for locating a water vein, he also developed a special technique for determining depth. He would hold a thin straight stick (five feet in length) over the water vein, and it would "involuntarily" nod up and down. The number of nods indicated the depth in feet to the water. During the same period thirty-two drillings were made in locations that were not dowsed; twenty-five resulted in successful wells and seven in dry holes.

In dowsing his own well, the water witch was over the Tertiary formation, and water (in small quantities) was located at 230 feet. In the case of the wells which the dowser "successfully" located, the wells were either drilled into the Tertiary formation or the farmers were willing to drill to greater depths which reached the Mesa Verde formation. The dry holes were cases in which the wells were either located in areas where the Tertiary sands did not exist and the farmer was unwilling to go deep enough to strike the Mesa Verde formation (two cases), or located over the Tertiary formation but in locations where because of buried channels or the presence of the syncline in the underlying Mesa Verde formation, the depth

HOPEFUL MASTERY OVER NATURE

to water was greater than the wells were drilled (three cases). The same geological facts account for the dry holes which resulted from drilling in locations that were not dowsed. The frequent errors in estimating depth were undoubtedly due to these same geological conditions, especially since the dowser usually named a depth which approximated or was less than the depth of his own well.

Not all the Homesteaders believe in and practice water dowsing. Our data indicate that opinions range from those of farmers who are wholly oriented in terms of the rational-technological methods of modern agricultural science and scoff at "those silly superstitions," to those of the farmers who believe firmly in, and practice, all three types of folk-ritual. One of the latter group is the water witch who is also full of knowledge about belief in "signs," "planting by the moon," etc. The data further indicate that of the three types of ritual, the water-witching pattern is the most widespread and the most persistent in spite of education in scientific methods. When the recorded instances of ritual practice from our field notes were classified, it was found that 57 per cent of the references were to water witching, 29 per cent were to the use of the almanac and the signs of the zodiac, and 14 per cent were to the use of natural phenomena to predict and control events.[29] It was further discovered that some of the most highly educated individuals in the community were having wells dowsed (for example, the principal of the school, who has an M.A. degree). Again, research revealed that opinion varied from utter skepticism ("the best witching stick is the end of the driller's bit") to complete faith in the ability of the dowser to locate water. The most frequent response is expressed in such statements as: "Well, I'm not sure I believe in it, but it don't cost any more"; or, "I'll always give it the benefit of the doubt."

Comparative data from a recent study in the Texas Panhandle are illuminating for the analysis of water witching.[30] The Panhandle study, initiated to provide controls over certain variables in the Homestead study, was focused on the small community of Cotton Center (population: 100), near the geographical center of the area from which the Homestead families came. The kinship and intervisiting ties between the two communities are extraordinarily close despite the distances involved, and there is ample evidence that the cultures of the two communities are still quite similar.

Specific inquiries as to the geology and ground-water resources and the use of water witching were made in Cotton Center. The community is located in Hale County, which is extremely flat and consists of slightly undulating hills interspersed with many poorly drained depressions that fill with water during the rainy season. Annual rainfall averages twenty-two inches, or almost twice the precipitation in Homestead. Most of the usable ground water is found in the Ogallala formation, a sandy deposit lying at or near the surface throughout the region.[31] The ground-water table stands at a depth of about 125 feet below the surface, and good wells can be obtained at almost any point. Wells are located where water can be used to best agricultural advantage on the farms. Water witching is widely known, but it is almost never practiced. It can be classified as an "unused skill" in Cotton Center.

There is also evidence from this area of the Panhandle to indicate that the practice of water witching in Homestead is not due to selective migration — that is, the "superstitious people" did not move west to New Mexico and leave behind the families with a more rational-technological orientation. For in Floyd and Crosby Counties, which are less than twenty-five miles to the southeast of Cotton Center, the ground-water situation is more variable; there is more difficulty locating wells; and water witching is currently practiced. Indeed, there are cases of men from Cotton Center who assist their relatives in dowsing for water in these other counties.

The question now arises as to why ritual patterns like water witching continue in spite of training in modern science. I have elsewhere examined the following three alternative theories which might account for the persistence of this ritual.[32] (*a*) The technological theory asserts that water witching is an empirically reliable technique for locating underground water. There is ample evidence both from Homestead and from other parts of the world to eliminate this theory as a possibility. (*b*) The survival theory is the basis on which many rural sociologists, government agricultural experts, and other observers of the rural scene dismiss folk rituals as "superstitions" which survive among unenlightened farmers who learned them from their fathers and grandfathers. This theory does not account for the fact that some of the more highly educated individuals in Homestead still resort to water witching, nor for the fact that although the educational level is approximately the same in the two communities, water witching is not used to locate wells in Cotton Center (where the water table stands at a uniform depth), but is a flourishing practice in Homestead (where the underground water supply is extremely variable in depth). (c) The functional theory views such practices as water witching not merely as "superstitious survivals" but as ritual responses to situations of technological uncertainty in the contemporary scene.

In the elaboration and application of this functional theory to the phenomenon of water witching, we must first specify the aspects of the Homestead situation which are technologically uncertain (and hence productive of emotional and cognitive frustration) from the point of view of modern science. It is clear from geological descriptions of the region that ground water is available in two of the formations (the Mesa Verde and the Tertiary sands and conglomerates) which underlie all or part of the region. One or the other of these aquifers can always be reached if wells are drilled deep enough. But it is equally clear that even with the most careful geological mapping, there exists a high degree of uncertainty as to the depth and amount of ground water available in any *particular* location where one may choose to drill a well. This factor of indeterminacy arises from the fact that surface outcroppings do not provide complete knowledge of the buried channels and ridges that resulted from erosion of the Mesa Verde formation in Tertiary times, nor of the structural conditions resulting from faults. Both of these geological factors result in substantial variations in the depth and quantity of ground water.

We have, then, a situation in which family-size wells were needed on the scattered farmsteads to relieve farmers of the time-consuming and expensive task of hauling water from some distance. Further, there existed a zone of indeterminacy in the exact location of adequate ground-water resources. The "stage" was set, so to speak, for the development of some kind of method to cope with the situation. Two things happened: on the one hand, a local farmer "discovered" that he had the power to dowse wells and began to do so throughout the community; secondly, geologists from the Soil Conservation Service began to visit the community and make certain recommendations for the location of wells. The alternative methods were in competition, but while the geologists came to the community infrequently [33] and could only provide answers to the question of location and the depth to water in a given well in *general* terms, the water witch was always available and could specify an *exact* location and an exact number of feet to water. These reassuring answers encouraged many Homesteaders to drill wells. When good wells were obtained at approximately the depth named by the water witch, his praises were sung throughout the community. When a dowsed well was a failure (because it was not deep enough), there were ready-made rationalizations to account for the failure.

In summary, there would appear to be a functional connection between the situation of technological uncertainty in locating wells in this region with an arid environment and complicated geological structure and the flourishing of the water-witching pattern. This conclusion is fortified by the observed fact that the area of greatest anxiety in the community — the location and development of adequate water resources — is also the area of the most persistent and most utilized ritual pattern, the water-witching technique (57 per cent of the observed instances of ritual practice); and by the fact that water witching is not practiced in the ancestral region of the Texas Panhandle, but is a common technique in Homestead.

Although a relationship can be demonstrated between technological uncertainty and ritual, the water-witching pattern has been elaborated and rationalized in terms of one of the central value-orientations of Homestead. For, in their relationships with the natural environment, the Homesteaders, strongly emphasize the orientation which may be described as "rational mastery over nature"; the environment is viewed as something to be controlled and exploited for man's material comfort. Thus, for the adherents of water witching in the community, the pattern becomes an important expression of this value stress upon "rational" environmental control. It is part of the farming process, along with clearing the land with bulldozers, plowing, planting, and cultivating the fields with power machinery, to locate a well by witching before one employs a driller. Explanations of how dowsing works are (predominantly) sought in terms of "electricity" and other "scientific" conceptions. Errors are attributed to the presence of metal objects which "short-circuit" the process, or to technologically faulty equipment.

There are clearly functional "gains" in the practice of water witching for the development and continuing survival of a community like Homestead. The certain answers provided by the dowser relieve the farmer's anxiety about the ground-water resources and

HOPEFUL MASTERY OVER NATURE

inspire confidence for the hard work of developing farms. The pattern also provides a cognitive orientation to the problem of why water is found (at a particular depth) on one farm and not on another — i.e., that there are water veins which run irregularly under the ground.

It is equally clear that the practice of dowsing involves certain functional "costs" in this situation. Energy and resources are invested in a technique which does not provide any better information as to the location of shallow underground water supplies than does the good judgment of individual farmers. In fact, dowsed wells are often located at spots which are extremely inconvenient and inefficient for the most economical operation of the farm. Some farm houses in the community are located in virtually inaccessible locations on the sides of hills "because that is where the water witch found the water." The Homesteaders who are adherents of dowsing believe that they are being "scientific" about locating underground water. This attitude tends to obstruct efforts to gain more precise geological information which, even if it does not tell the farmer the exact number of feet to the water level at a given location, is at least a more promising long-range approach to the development of water resources for the community.

THE "MOON MEN"

In order to focus more clearly the analysis and interpretation of ritual patterns of control as developed in Homestead, I have described the water-witching phenomenon at some length. It should be emphasized that these sections on ritual methods of "control" over nature are given more extensive treatment than the rational-empirical methods not because they loom larger in the total life of the community (for they do not), but because it is assumed that the reader will have greater familiarity with the techniques of empirical control than with the techniques and functions of the ritual patterns. In addition to the "witch men" (as the proponents of water witching are called), there are also "moon men" and "sign men."

The "moon men" are those who utilize phases of the moon to predict weather and to guide their agricultural activities. The most important beliefs are that root crops should be planted in the "dark" of the moon (or during a waning moon), while above-ground crops should be planted in the "light" of the moon (or during a waxing moon). The phase of the moon also guides the leading ritualists in deciding when to plant in order "to get by the late spring frost." They say that "it will freeze late if a new moon comes before the twentieth of June, because when it's new moon and it's dark all night, then it really gets cold" (and, therefore, a late planting is advised). Other related beliefs include the notion that the butchering of hogs should be done during a waxing moon so "the lard will be good," and that cattle which are bred after the sun has passed the zenith (noon) are more likely to yield heifer calves.

THE "SIGN MEN"

The "sign men" are those who utilize the signs of the zodiac printed in the almanacs,[34] which are distributed free in the local store to guide certain agricultural and livestock activities. The most common beliefs are

You should castrate when the signs are around the legs, so the animals won't bleed to death.

You should dehorn cattle when the sign is in the heel to prevent bleeding.

You should wean calves when the sign is in the knee, so they won't get thin and bawl around.[35]

You should butcher when the sign is in the heart, so the meat won't be tough.

Although the use of the moon and signs to guide farming activities is less widely believed and less practiced than water witching, it is significant that almost all of the beliefs are focused around the areas of uncertainty and anxiety in farming life. These rituals, like water witching, are also regarded by their proponents as "scientific" procedures; the practitioners believe that the phases of the moon or the signs of the zodiac have a naturalistic connection of some kind to the weather, the crops, and the livestock. While the techniques may be regarded by the scientific observer as nonempirical means for achieving empirical ends — hence, functionally equivalent to the magical practices of nonliterate societies — the adherents in Homestead regard such practices as rational-empirical techniques; hence, they are part of the general value-orientation, "mastery or control over nature." The techniques can, therefore, be best described as a type of "folk science" or "pseudo-science" which

HOPEFUL MASTERY OVER NATURE 87

is present in the local cultural tradition. As a body of pseudo-scientific knowledge, these patterns in rural farming culture are the same type of phenomenon as the pseudo-scientific practices which center around situations of uncertainty in other areas of our culture. For example, in modern medical practice there appears to be a pattern of "fashion change" in the use of certain drugs, an irrational "bias" in favor of active surgical intervention in doubtful cases, and a general "optimistic bias" in favor of the soundness of ideas and efficacy of procedures which bolster self-confidence in uncertain medical situations.[36]

PRAYER MEETINGS

A completely different kind of magical ritual is the manipulation of formal Christian religious symbols to achieve empirical ends. The means used are admittedly not "rational" or "scientific"; instead, the Homesteaders believe the techniques depend upon "supernatural" power, and one has to have "faith" to make them work. A typical example is the "old-time prayer meeting for rain." These meetings are found predominantly among the more fundamentalist Southern denominations and were held quite frequently when the Homesteaders lived in Texas.[37] An older Homesteader, recalling one of them, said:

> I remember a prayer meeting in Bosque County when I was a small kid. It was during a bad drought. They were meeting and praying for rain all over the state. This particular day they met in the school house, and there were no clouds in the sky that morning. They started praying for rain about 11:00 A.M., and about 2:00 P.M. it was raining until you couldn't see.

In Homestead most people now express skepticism about these prayer meetings for rain. Only one such meeting has taken place in the history of the community:

> They had a Baptist revival here in 1934 and we needed rain bad. I remember we divided up. The men in one circle and the women in another circle. We got all ringed up and each one was supposed to say a prayer. I remember how the preacher said he had never had a prayer meeting without it raining, but they had a prayer meeting for a whole week, and we didn't get no rain.

A number of the older people, however, still think that "prayer meetings would work if you had enough faith to make it work, but there's not enough faith here in Homestead." Many people also say prayers privately for rain during the critical dry season.

A related conception is that "righteous living" has an effect upon rainfall,[38] but on this point there is a basic split in opinion between the more fundamentalist Baptists in the community and the Presbyterians (the two most important churches). The Baptists believe that "good" people are favored, and they quote from Isaiah to justify this position: "I will also command the clouds that they rain no rain upon it [the vineyard]" (Isaiah 5:6). Also, the following statements are from a Baptist sermon:

> Righteous people get rains for their crops, but drunkards do not. I have often seen rains stop at a barbed wire fence. One time in southeastern New Mexico I used to know two men who lived side by side. One was a righteous man and he had good rains and good crops year after year. But the drunken reprobate who lived just across the fence got no rain and no crops.

The Presbyterians, on the other hand, quote from St. Matthew to the effect that the Lord "sendeth rain on the just and the unjust" (Matthew 5:45).

FAITH HEALING

The notion that certain people are endowed with "supernatural" power derived from God to effect cures of various kinds was also common in the earlier cultural background of the Homesteaders.[39] These "faith healers" could "take off warts, stop bleeding, cure a sick headache, and cure animals of worms."

Although there have never been any faith healers in Homestead, at least two Homesteader families have made trips to see Mrs. Susie Jessel, "the Miracle Woman," at Ashland, Oregon.[40] Mrs. Jessel was born in the mountains of Tennessee, and has been a practicing healer since she was sixteen when she "received a vision in her father's plum grove." She is reported to give treatments to from 400 to 600 people a week and charges nothing for her efforts, but accepts small coins that "are dropped into her apron pocket." Mrs. Jessel says that "there is nothing miraculous about my work; I simply use whatever power my Creator has endowed me with." She works by passing her hands over the body of the patient, and the infirmities and aches of the patient are supposed to leave the patient and enter into the hands of the healer as she works.

The veins on the backs of her slender hands raised up thick and full in difficult cases; they were almost normal in easy ones. On occasion, stubborn diseases raised her arm veins nearly to the elbow and her hands became hard as rock. At the touch of her water-saturated towel, however, these body accumulations and swellings apparently disappeared.[41]

The following diseases have reportedly been cured by Mrs. Jessel: cancer, tumor, goiter, arthritis, appendicitis, abcesses, asthma, yellow jaundice, sleeping sickness, stomach ulcers, paralysis of limbs, adhesions, and cataracts.[42]

One of the women from Homestead who went to Mrs. Jessel for a "cure" was suffering from "glandular trouble" which supposedly was making her excessively fat. Although she was not cured of this ailment, she still has faith in the healer because she was cured of another ailment: "her arms used to go to sleep while she was in bed" and they no longer do so since her visit to the healer. The second family made the trip with the hope of finding a cure for an alcoholic brother-in-law. It is reported that "he was improving for a time under Mrs. Jessel's care, but then he went on a big drunk and died."

"TALK ABOUT RAIN"

It has, of course, been repeatedly observed that people all over the world talk about the weather in making polite conversation. But the constant talk about rain throughout the farming season in Homestead is more than polite conversation. In this case there is no conception that this talk will bring rain in any "naturalistic" or "supernaturalistic" manner. Rather, the conversation appears to be merely an "expressive" orientation to a central anxiety, as if the constant use of the word symbol "rain" makes people feel better even when they think they cannot do anything about it.

The importance of this type of orientation in Homestead derives from the fact that it is shared by all members of the community. Some people believe in rain making with silver iodide, others do not; some think that perhaps prayer meetings might help, but others scoff at the idea. Yet everyone talks about rain, and on any given day during the summer, small groups of Homesteaders gather in the stores, the shops, or the bar, all wondering when the next rain will come or telling long stories of previous rainy summers. Often the tension is expressed in humorous remarks,[43] such as, "Now I know why this country is full of old Indian ruins"; or, "Sometimes drought is a good thing for a community, because it gives people something to talk about besides one another." The atmosphere is always laden with tension and anxiety until the summer rains arrive and the bean crop is no longer in danger of failure. The constant talk about rain seems, to some extent, to alleviate the tensions and anxiety; reassurance that rain will come again is given by recitals of previous heavy summer rains.[44]

THE GAMBLING ORIENTATION

I have described the rational-empirical methods, the "pseudo-scientific" techniques, and the "supernatural" procedures through which the Homesteaders hopefully try to cope with and control the processes of nature. Although the Homesteader equips himself with the most modern type of tractor, watches the moon to help get by the late spring frost, and even says prayers (public or private) for rain, he still realizes that there are uncertainties about farming in this semiarid environment that extend beyond all his attempts to master the situation. At this point, the Homesteader has a strong tendency to assume a gambling attitude toward the whole enterprise. The Homesteaders express this attitude explicitly with such statements as: "It's good luck when we hit and bad luck when we don't"; or, "We're betting our labor, our seed, and our fuel against whether we make anything."

Unlike the neighboring Spanish-Americans, who believe in absolute chance (*Sea lo que sea* — "what will be, will be"), the Homesteader believes in a chance which can be outguessed and outmaneuvered.

If the Spanish-American is like the gambler who bets blind — who takes what he is dealt as inevitable — the [Homesteaders] are more like the gambler who peeks at the corner of his card, eyes all the other players, struggles over whether to get in or stay out (usually getting in and regretting it) and, what is worse, finds himself, a dry farmer in an arid land, continually drawing to an inside straight.[45]

Like the chronic gambler, who is at times pessimistic but usually feels that he may win in the next game, the Homesteader maintains enough confidence in his techniques and his "luck" to go on trying. "We always think that if we don't make it this year, we may make it next year."

HOPEFUL MASTERY OVER NATURE   91

This gambling orientation is also characteristic of farmers on the Great Plains. Bell reports:

> The variability of the rainfall has given the people a feeling that luck is more influential than most practices. This attitude is reflected in their entire personality organization . . . It has helped to develop the gambler's psychology, so noticeable to the outsider . . . They have an optimism which is nearly unbelievable. Perhaps it is rivaled only by the race track follower who is always sure he will win next time.[46]

Bell further describes this gambling psychology as a response which emerged when farmers accustomed to farming in the more humid areas to the east moved onto the Plains and were confronted with environmental conditions which proved to be beyond the control of their familiar farming methods.

Thus, the Homesteaders maintain a gambling orientation which emerged back on the Plains and was reinforced by the hazardous farming conditions in their new settlement in western New Mexico. With their Texan background (for the most part) the Homesteaders also maintain what might be called a culturally patterned "boasting psychology" which is related to their gambling orientation. The tendency of the Texans to "boast," "exaggerate," and "tell tall tales" is famous the world over. This boasting pattern would appear to be an aspect of the generalized American "love for bigness,"[47] which has been especially over-emphasized by Texans, who inhabit "the biggest state in the union," own "the biggest cattle ranches," and grow "the biggest crops."

Like the man who whistles in the dark before the storm, the Homesteader attempts to master the situation in a "big" and "boastful" fashion. He buys a new set of farm machinery and plants the best bean seed and boasts about how big a crop he is going to make. Then, if frost or drought destroys the crop, he boasts about his losses like a big-time gambler. He either wins in a big way or loses in a big way; and the next year he tries to outguess and outmaneuver the situation once again.

CONCLUSIONS

In this chapter we have attempted to trace the influence of the value-orientation of "hopeful mastery over nature" upon the course of events in Homestead. Today the Homesteader confronts the natural environment with a set of twentieth-century tools and applies this equipment to its greatest capacity in his endeavor to subdue and control nature toward the achievement of his own ends. It is also clear that with this orientation much has been accomplished in a practical, realistic way. Homestead, with its large fields, wide roads, windmills, stores, and large bean warehouse, contrasts markedly with the settlements of Navaho hogans in the pinyon trees or with the rock and adobe Spanish-American villages which blend timelessly into the landscape. The stress upon the value-orientation of control over nature has (within the inevitable limits imposed by the physical environment) influenced the *selection* of ways of dealing with nature (especially in terms of a high degree of mechanization); it has continued to *regulate* the man-environment relationship (especially by providing cultural definitions which prescribe the desirability of trying to control natural conditions rather than adopting a fatalistic approach); and it has provided a *ranking of goals* by holding out the promise of greater prestige to the Homesteaders who acquire new and more powerful machinery and attempt to improve their farming methods.

The Homesteaders have learned from continuing experience with an uncertain agricultural environment (on the Great Plains they were already beyond the twenty-inch annual rainfall line) that there is a limit to the efficacy of rational-empirical methods and "ritual" techniques in the control of the natural situation. Nevertheless, the hope of mastering the natural environment persists, and a "gambling psychology" and associated "boasting pattern" have developed which come into operation when rational and ritual methods fail. Continued optimism in the face of repeated failure is related to the future-time orientation discussed in the next chapter.

4

Living in the Future

To look forward to the future, to forget or even reject the past, and to regard the present only as a step along the road to the future, is a cherished value in American culture and a conspicuous feature of life on the frontier. This future-time orientation and associated value emphases on "progress," "optimism," and "success" have had a profound influence on the settlement and development of Homestead.

When asked whether the people of Homestead "live" in the past, present, or future, one of the original settlers replied:

> Why, I'd say we live in the future. We're always looking forward to the future. Once in a while some of us gets together uptown and talks about the past, but everybody is for the future. What's done past, I don't care a thing about that.

When the Homesteaders left the Plains and moved westward "to better themselves," they were thinking of their future, for the frontier to the west of the continental divide held the promise of free land and permanent homes. Indeed, the early conceptions of the land they were settling were grossly exaggerated; the Homestead area was commonly called a modern "Garden of Eden." One of the first "preachers" to reach the new community had had a "vision" of what he would find: a region with fertile land, lush grass, and where tall corn and large cabbages could be grown!

Gradually this conception of the nature of the land has been tempered by realistic experience, but the Homesteaders have not changed in their tendency to live for the future. Certainly there is some talk about the "past" in Texas or in the early days of Homestead, but the past is not glorified or called upon as much to provide a rationale and justification for the present as it is by the Pueblos, who draw upon their myths as guides for present-day behavior. Other Homesteaders tend to live from day to day without much concern for either the past or the future, and in this respect they approximate the orientation of the Navahos and Spanish-Americans. But most people in Homestead are very explicit in word and deed about "living for the future." Even in the face of a severe two-year drought in 1950 and 1951, most Homesteaders expressed the firm belief that Homestead would nevertheless grow and prosper.

"PROGRESS"

One of the most important aspects of the future-time orientation is the notion of "progress," which has also been emphasized since the earliest formation of American culture,[1] particularly on the frontier where the pioneers had to "start from scratch" in the process of building farms, homes, and communities. As a value-goal, "progress" has been a dynamic element in Homestead since the early days of settlement. There has been a persistent striving toward "bettering" the "places" on which the Homesteaders live, together with a striving to make Homestead the "Pinto Bean Capital of the World." The measure of this "progress" has always been economic and material, rather than social or spiritual. It has meant primarily "bigger and better" farm machinery, houses, stores, roads, etc., rather than progress in better relationships among people in the community or with neighboring groups such as the Spanish-Americans.

Progress on the Homesteaders' "places" has involved an increase in the acreages under cultivation from "small patches in the woods that were farmed when we first come here" to the present 400 to 600 acres on many of the farms. It has involved a shift from the use of horse-drawn implements and the "old thrashing machine" to the modern tractor and combine. Further, it has meant a change from small subsistence farming to large-scale commercial farming.

In their farm houses, the Homesteaders have tried to change from the "old log cabin" with kerosene lights to the "modern" adobe or frame house with electricity and "all the conveniences." In the words of one of the women:

> We always look forward to getting things we didn't have, like electricity, ice box, deep freeze, new couches and kitchen cabinets.

These "modern" house-types vary in size from three to six rooms; some contain minimum furnishings, and at the other extreme a few are equipped with modern furniture, bathrooms, and items charac-

LIVING IN THE FUTURE   **95**

teristic of "middle-class" homes in urban areas. Nine of the sixty-one homes in the community now have running water in the house, and six of these nine also have bathrooms. The other farmsteads depend upon water carried from a well on the farm (or hauled in barrels from the community well) and upon outhouses. Wood is generally used for heating and cooking, but a few families have butane and electric cook stoves; others have kerosene heaters, and two homes are heated with butane. One half of the homes are now wired for electricity; the others are out of reach of the present power line, but most will be included by lines now under construction.

Today the Homesteaders look forward to the time when they will all have electricity, running water, indoor plumbing, "and everything modern." The electrician who wired the houses for electricity reported that, "Every family has some little old room which they call a bathroom or hope will be a bathroom some day." When the first bathtub in Homestead was installed in 1950, the people said, "It'll be just like living in town." The Homesteaders feel they have the "highest standards of living" of any of the cultural groups in the area, with the exception of the Mormons in Rimrock, and that they want "to keep on making progress." There have always been a few Homesteaders who make little or no effort "to keep up with the times," and the families who still live in old houses and log cabins are subjected to constant criticism by the rest of the community.

The Homesteaders felt when they left behind the "more civilized" communities in Texas and Oklahoma and moved to the frontier that they were settling in "a jumping-off place" at the western edge of their familiar world, but they also believed strongly that in due time the town would grow and prosper. The Homesteaders have always been the most conspicuous "boosters" in western New Mexico. The following article, written by the chairman of the Boosters Club for the *Homestead Promoter*, epitomizes the Homesteader's conception of his town:

HOMESTEAD – PINTO BEAN CAPITAL OF WORLD [2]

 The famous Homestead pinto beans entitle Homestead to be called the new "Pinto Bean Capital of the World." The quality of Homestead's pinto beans brings Homestead into the limelight of the bean growing world. There seems to be no more suitable soil nor more suitable climate for the production of an abundant crop and a superior grade of pinto beans in any dry land farming area known to the pinto bean growing world than that of the Homestead area. The famous Homestead pinto bean becomes King and promises to open up a greater market by leading to a greater consumption of pinto beans. They grow bigger, cook easier, and taste better. Growers in the Homestead area are urged to put forth every effort possible to increase their pinto bean acreage in order to supply the ever growing demand for the famous pinto.[3]

The Homesteaders are convinced that their town will become "bigger and better," and that in time it will be transformed from the group of scattered farms into a modern city "like Plainview, Texas," with paved streets, several stores, a movie theater, a hospital, churches, a fine modern school.

The town started in 1932 at an intersection of two of the section-line roads. One of the families occupying land at this crossroad started a store. The families on the other three sections began to divide portions of their homesteads into lots which were bought and sold in a brisk market. Within a few years the "essential" institutions were established: a school, a post office, stores, a blacksmith shop, and a *café*. At the time of the largest population in the 1930's there was even a hospital (for one year) with a doctor (who was an osteopath) and a nurse (the doctor's wife), as well as a lumber yard, a real estate agency, a soda fountain, and a "Little Theater Group."

The Boosters Club was organized to promote the growth of the community; in addition to publishing articles in the *Homestead Promoter*, the club sponsored dinners to which business leaders from nearby cities were invited, who came to eat the products grown in Homestead and to give speeches about the progress the community was making. A baseball team was organized, which traveled over three counties winning games and "putting Homestead on the map."

By 1935 the population had reached a peak of 374. In 1940 it declined to 333, and many of the "speculators" and "drifters" were gone. For, although the bulk of the population was composed of "solid and respectable" families seeking homes on the frontier,[4] it was clear that Homestead, like all pioneer communities, also provided an opportunity for speculators to make a quick dollar and was a refuge for others who were "just one jump ahead of the law." As a refuge from the law, Homestead was the home of a colorful succession of individuals who arrived, stayed a year or two, and left again; but while they were in town, there were bootleg stills in the woods, people living on ranchers' beef, and there was more than an ordinary

LIVING IN THE FUTURE 97

amount of "stealin', drinkin', cussin', and fightin'." Although Homestead continues to be a haven for an occasional fugitive and all the families who made bootleg liquor are not gone, the population since 1940 has been composed mainly of a solid core of Homesteaders who are determined to stay and build permanent homes.

Despite the fact that the population of the community has been reduced by almost half (in 1950 it was 232) and many places of business and organizations are no longer active, the "boosterism" still goes on. There was general rejoicing when the power line, which has become an important symbol of progress, reached Homestead in 1949. One of the original settlers announced:

> This community is a proven fact now. It makes me feel good every time I ride along on my tractor and look over and see the poles of the power line coming into Homestead. Someday we'll get a road too, and then this community will really be something.

In 1950 plans were initiated to build a large and elaborate highschool gymnasium. In response to pressure from the school principal and one of the many informal committees that are always organized to promote things, the state appropriated $10,000 to be used for materials, providing the Homesteaders would donate their labor for the making of adobes and the construction of the building. At this point, the value placed upon individual interests took precedence over "progress," and the plan was rejected by the Homesteaders with speeches to the effect that "I've got to look after my own farm and my own family first; I can't be up here in town building a gymnasium." Later the same year additional state funds were appropriated for labor; and with these funds adobe bricks were made, the foundation was dug, and construction was started — the Homesteaders had agreed to work on the gymnasium on a purely "business" basis, at $1 an hour. As soon as the funds were exhausted, however, construction stopped. At present, a partially completed gymnasium (and stacks of some 10,000 adobe bricks disintegrating slowly in the rains) stands as another monument to the rugged individualism of the Homesteaders.[5]

After this temporary setback, the "march of progress" continued in Homestead. Encouraged by the arrival of the power line, the Homestead Farm Bureau [6] (whose functions are mainly centered around the sale of automobile insurance and the promotion of community enterprises) appointed a committee to investigate the possibilities of a telephone line; letters were written to various departments in Washington.

Renewed efforts were also made to obtain a paved highway through Homestead. Since the early days of settlement, "road committees" have been appointed periodically and sent to the county seat and state capitol to ask the authorities for road improvements. To "keep living in the mud" during the summer rains and winter snows has always been an intolerable situation for the Homesteaders; in contrast, the neighboring Spanish-Americans and Navahos have always taken for granted that one must use and be satisfied with rural dirt roads. The Homesteaders have been convinced that it would only be "a year or two" before a paved highway would pass through Homestead. They have calculated that "when a paved highway is built direct from Albuquerque to Phoenix, it will go right past Homestead"; and "when they build a road from Salt Lake City to El Paso, it will go right through Homestead." These ideas have long been cherished in the Homesteaders' dreams of progress, but the state highway engineers of the various states have never planned such roads, and there is no economic or geographical reason why they should be constructed. Nevertheless the "road committees" have continued to put constant pressure on the authorities. Occasionally, graders were sent from the county seat to level off some of the bumps in the roads, and small stretches of road were graveled in election year to appease the voters. However, the state politicians had not anticipated the persistence and stubbornness of the "progressive" Homesteaders. Eventually a road grader was permanently stationed in the community.

In the fall of 1951, a large "road meeting" was held in Homestead. Invitations were sent to various business and political leaders of the surrounding cities, many of whom appeared and made promises to help with the road situation. This was followed by two later meetings in Gallup which were attended by the state highway engineer and members of the New Mexico State Highway Commission. Later the county agricultural agents were asked to collect data on the use and need for roads, and the road from Homestead leading north toward Gallup was given first priority in the use of secondary-road funds alloted to the county by the federal government under the "farm-to-market" road program. In the autumn of 1952, surveying

LIVING IN THE FUTURE

crews finally appeared and started planning a road program, and the Homesteaders said: "We can see a paved road right now, running right down through Homestead." They were again far ahead of the road builders in their dreams for the future. The Homesteaders were planning tourist courts, *cafés*, and filling stations to take advantage of the tourists who, they thought, would soon be pouring through Homestead.

"OPTIMISM"

Closely related to the Homesteaders' emphasis on "progress" is an orientation of "optimism" toward the future which, despite a few pessimists in the community, is the predominant outlook on life in Homestead. Optimism is apparent both in the hopes for progress cherished by the Homesteaders and in their gambling orientation toward farming. As previously shown, the concepts of progress involve a perennial optimism about the improvement of material conditions on the Homestead farms and the steady growth and eventual prosperity of the community; the optimism transcends the facts of population decline, periodic drought, and crop failures.

In the midst of the severe drought of 1951 (when only twenty-five sacks of beans were harvested in the entire community and logically the Homesteaders might have pulled up stakes and moved on), one new house was built and three were improved by the installation of running water. Each year the wife of the owner of this new house had expected that her new home would be built; when it finally was, people said: "She waited eighteen years for her house, but you see she finally got it." Even the one "squatter," who for sixteen years has fought the efforts of the Bureau of Land Management to put him off the land, is convinced that eventually he will be given an opportunity to file on a homestead, and he continues to plant and harvest beans each year.

The gambling orientation toward the farming enterprise demands, of course, strong optimism; the bean farmer must always feel that "even if he didn't make it last year, he will make it this year." One of the bankers in a nearby city commented that the Homesteaders "are the greatest next-year people in the country." He described how the bean farmer comes to town each winter, reports that he cannot pay off his loan, and asks for another year's extension because he is "sure he'll hit a crop next year."

"SUCCESS IN HOMESTEAD"

The owner of a "place" (farm) in Homestead is in competition with all the other families who have "places"; this pattern is the familiar American one of striving to "keep up with the neighbors." The emphasis on individual "success" is an aspect of the future-time orientation, for further or ultimate success always awaits attainment in the elusive future. In Homestead the conceptions of success revolve primarily around economic achievement: the expansion and improvement of one's land holdings, the acquisition of newer and better tractors, automobiles, home appliances, and gadgets, and depends upon and includes the production of bigger and better bean crops. There are, however, no absolute criteria by which the Homesteader measures supreme success; instead he compares his possessions and achievements with those of his neighbors and tries to keep up with them.

Unlike the findings of Warner and others [7] that middle- and upper-class families are reluctant to discuss their incomes frankly, the Homesteader is extremely free with information not only on the amount of money he makes, but on how much he paid for land, for furniture or objects for the house, for tractors, cars, etc. A detailed, endless discussion of a recent transaction will ensue from the question: "How much did you give fer it?" The game of matching accomplishments and comparing possessions is made considerably easier by this straightforward approach .

Each year the farmer looks forward to harvest time when the sacks of beans will be weighed and counted and he will learn how successful he has been. Enormous prestige is accorded those farmers who have succeeded in producing the largest total amount of beans and to those who have produced the largest yield per acre. Early in the farming season the "loafing groups" which gather uptown begin to assess and compare the crop prospects of the various farmers. These sessions continue throughout the summer, and after harvest the box score is known to everyone; the "top bean men" are envied and talked about during the winter. Next season the process begins again, with the largest producers looking forward to maintaining their positions or bettering them, and others striving to "go them one better."

The "failures" in Homestead are the farmers who fail to make a crop even in good farming years and are those who continue to occupy positions of hired men or tenant farmers and make little effort to acquire "places" of their own. It is expected that anyone may be a failure in a poor farming year, but a half dozen or so farmers are singled out for special comment because "they have never made a go of anything."

The women for the most part derive their prestige in the success scale from the economic achievements of their farmer-husbands. In four cases, however (two schoolteachers, the postmistress, and a woman who does most of the farming for her family), the women hold important economic positions and contribute significantly to the success of their families in the community.

"SUCCESS IN THE OUTSIDE WORLD"

It is relevant to examine what has happened to the people who have left Homestead over the years — where they have gone and how successful they have been (a tabulation, by family, of this emigration is found in Table II). Although the immigrants to Homestead

TABLE II

EMIGRATION FROM HOMESTEAD BY FAMILY

Place of immigration	Farm owners	Farm laborers or tenants	Non agricultural	Total families
Texas (South Plains)	5	2	16	23
Middle Rio Grande Valley	8	0	11	19
Pacific Coast	1	8	6	15
Eastern New Mex. Plains	9	0	2	11
Arizona	0	1	4	5
Gallup and Grants, New Mex.	0	0	5	5
San Juan Valley, New Mex.	2	0	0	2
Southern New Mexico (Deming region)	2	0	0	2
Colorado (Delta region)	2	0	0	2
Missouri	1	0	1	2
Idaho	1	0	0	1
Unknown*				20
TOTALS	31	11	45	107

* No information exists as to where the twenty families listed as "unknown" have moved. The Homesteaders are no longer in touch with them.

came from a single geographical area, it is evident that there has been substantially more "scatter" in the emigration from the community. The greatest number have gone eastward, back to the Rio Grande Valley and the Plains of eastern New Mexico and Texas. This movement represents well over 50 per cent of the known cases. Slightly less than 25 per cent of the known cases have gone west to Arizona and the West Coast, and the remaining 25 per cent are scattered for the most part in New Mexico and other intermontane states.

As to the present occupations of the emigrants, forty-two families (or almost half the known cases) are still engaged in agriculture, and nearly three-fourths of these families now own small farms. Significantly, only one of the nine families engaged in agriculture on the Pacific Coast (where large-scale commercial farming tends to dominate the agricultural scene) owns a farm. Elsewhere, especially in New Mexico and Texas, almost all of the emigrant agricultural families have became farm owners. The other half of the emigrating families are engaged in various nonagricultural occupations, such as laboring jobs (predominantly unskilled and semi-skilled labor with the exception of some skilled plumbers, carpenters, and welders) and clerical positions. Only six families now have small businesses of their own.

It is significant to note that this migration away from Homestead has also been a highly selective one, for it is clear that the geographic location and present occupations of the emigrants place them within the area populated by people of their own kind — Texans and Oklahomans. Further, the bulk of the emigrants have not followed the stream of migration westward to the Pacific Coast (where it is difficult to purchase a small farm) but have tended to move to those areas where they could maintain continuity of patterns and values.

Many young people have also left Homestead; during World War II selective service (and voluntary war service) took an extraordinarily large number from the community. In all, forty-six young men and five young women (or approximately 15 per cent of the population) entered the armed forces. Only one became an officer; the rest were enlisted personnel, which undoubtedly reflects the low educational level of the Homestead servicemen, who had only grade or high school educations. After the war only ten veterans returned to live in Homestead; the others moved to other communities. A Veterans Administration Farm-Training Class was organized in Homestead and was attended by all of the veterans as long as their

LIVING IN THE FUTURE

eligibility lasted. Despite this training, however, four of the veterans were not satisfied or were not successful as farmers, and today only six of the veterans still live in Homestead.

As indicated in Chapter 1, of the fifty graduates of Homestead's high school since its establishment in 1937–1938, only five have remained in the community. Of these five, all except one boy (who is still unmarried) have married within the community. The three boys are farming and the two girls are wives of farmers. Of the twenty young men and twenty-six young women [8] who have left Homestead since 1937, only twelve married within the community before they left. Seven of those who have left (four boys and three girls) attended college, and four (two boys and two girls) of those seven completed their A.B. degrees. Most of those who left the community are now in skilled, semiskilled, or clerical jobs or, in the case of the girls, have married men in these types of occupations (see Table III).

TABLE III
Occupations of Young People Who Have Left Homestead[*]

Professional		6
Electrical engineer	1	
Animal husbandry	1	
College professor	1	
Pharmacist	1	
High-school teacher	2	
Army-Navy enlisted men		5
Small farmer		4
Office clerk		4
Telephone linesman		3
Telephone operator		2
Plumber		2
Truck driver		2
Carpenter		1
Brick mason		1
Miner		1
Service station operator		1
Car salesman		1
Dog catcher		1
Maintenance man (street repair)		1
Common labor		3
Unknown		8
TOTAL		46

[*] Including Homesteaders' spouses from other communities.

Although Homestead has not yet produced any "big businessmen," "prominent politicians," or "millionaires," a few of the young people who have left the community are always described as "real successes." The most important is the young woman who was first a schoolteacher in rural schools in the area but continued her education and eventually obtained a Ph.D. in botany at the University of Iowa. She is now the chairman of the Department of Biology in one of the state universities. Another Homestead girl finished her A.B. degree at the University of New Mexico, where she met and married an electrical engineer. She now lives in New York with her husband. A third girl went to business school in Albuquerque after her graduation from high school, and during the war joined the Waves. She married a Navy petty officer and they now live in a suburb of Philadelphia where her husband is a pharmacist and owns a drug store.

One of the young men finished his A.B. in animal husbandry at the New Mexico State College on the "GI Bill" after the war. He now has a job on a large ranch where he provides the technical advice for raising registered cattle and for preparing "prize bulls" for competition in state fairs. He plans to attend a university and study for a Master's degree. Finally, one of the young men is now a high-school teacher, and another young woman is married to a high-school teacher. These high-school teaching positions are not considered as outstanding as the other "successes," but are nevertheless given more consideration than the group who hold laboring or clerical positions.

All of these "success stories" have become important legends for Homestead, and the cases are discussed again and again as examples of how it is possible for Homesteaders to "go out into the world and be a success." At the other end of the scale are those who hold labor jobs with no particular skill or security involved. These people are not often discussed, but when one does ask about them, the Homesteaders make it clear that these young people are considered "pretty much failures."

Although only a small percentage of Homestead's high-school graduates go on to college, the level of aspiration among the current high-school students is relatively high. In the autobiographies written by the sixteen high-school students in 1950 (in which they described their life careers up to the present time and then projected their

LIVING IN THE FUTURE

careers to 2000 A.D.), twelve of the students (six girls and six boys) hoped to attend college when they finished high school, and twelve hoped to leave Homestead for careers in the outside world. These expectations were phrased as follows:

High School Boys

(1) What I hope will happen in the future is that I will get to go to college and then when I get out, have a large ranch. I hope that Homestead will grow into a thriving town in the future and soon become a large town.

(2) I hope I get to go to college and go four years of agriculture, and when I get out of college I am going to play baseball for maybe five years and then I am going to build me a dairy in Colorado near Trinidad.

(3) I hope when I get out of high school I can go to college and learn about agriculture and science work. I hope that after I get out of college I can get a good job where I can buy a nice home, car and raise a good family . . . I expect to see the day when a highway comes through Homestead. I expect to see the community grow when it does. I expect to see a very good school and gymnasium here . . . and more modern homes here, and water piped into every home and bigger and better built stores and filling stations.

(4) I hope to work for someone besides my uncle this summer and save my money so that I can go to college and learn to be a veterinarian and stay in the community and buy me a small farm and get married.

(5) In the future I hope that everything goes well and that I decide to go on to college, but I expect that I will leave Homestead and get a job somewhere and work for wages the rest of my life.

(6) I want to travel over the Al-Can Highway to Alaska. I believe I would like Alaska because it is called 'America's Last Frontier,' and I have always liked frontier life.

(7) I want to be a great baseball player and after a few years I want to live on a farm and have a few cows and have a dairy. I imagine I will leave this community after school is out so I can go to college. But after I finish college I might come back here.

(8) I hope I finish high school in two more years and then I guess I will stay and help Daddy a year or two and then go out and see the world. Then I will start making a ranch of my own because I already have 30 head of cattle.

High School Girls

(1) I hope I will get to go to college, go to aviation school, and learn all about airplanes . . . I want to go to Texas to college. Then I hope to go to some flying school farther up north . . . I want my husband

to be tall, dark, and handsome, with plenty of money, but would be satisfied if he wasn't rich.

(2) I would like very much to go to at least four years to college, and I think the only thing which will stop me from going is 'money.' After I finish college, I want to be either a nurse, a bookkeeper, or a librarian. I want to marry and have children later on.

(3) I hope that Homestead will be a fair-sized town. Not as large as Albuquerque or even Gallup but about the size of Kingsville, consisting of several department stores and wearing apparel shops, movie theaters, gas stations, cafes, and so on. I think I'll be married someday, but I only want one child. I am going to college when I graduate, and I'd like to be a stenographer or a telephone operator.

(4) I hope I will pass every year in high school, then go four years to the University at Albuquerque. After I finish there I want to be a secretary of some big business outfit. Along with being a secretary I hope to be a pianist as I am taking lessons so I hope to get to be very good as the years go by. I hope I don't meet a nice young man until I'm at least 26 years old.

(5) When I finish high school, I want to go to college. I want to become an airplane stewardess. If I can't be a stewardess for any reason I would like to be a nurse or a secretary. Later I want to be a housewife and live on a ranch in Wyoming.

(6) As soon as I get out of school, I want to go to two years of college and then I want to enter nursing school. During my student nursing I'll meet my future husband. He will be a doctor. We will live in Riverside, California. I will not live in Homestead. I just hate it here anyway.

(7) In the future I want to go ahead and finish school. I then want to get a job to help save money toward marriage. I hope to get married when I am 19 or 20 years old. I want a nice home in the country. I would like most to live on a small but prosperous farm. I like to raise chickens.

(8) When I graduate, I may go to college, but I seriously doubt that. I also hope to live somewhere besides Homestead, New Mexico, maybe in Texas somewhere. I hope in my future to marry a certain guy. I want to learn to cook and sew, so I will make a good housewife.

The aspirations continue in each generation of young Homesteaders. Although only a few of those who hope to will actually go to college (according to previous trends), almost all will leave Homestead for careers in the outside world. When the present generation of farm operators becomes too old, it is possible that there will then be more room for younger farmers in the community. It is too soon to tell what will happen when the older farmers die and

LIVING IN THE FUTURE 107

the land is inherited by the younger generation (since most of the Homesteaders settled here when they were in their twenties or early thirties and are therefore still vigorous), but in the meantime the local schoolteachers encourage the young to go on to college and to be "ambitious" and "successful," while the "preachers" have a tendency to emphasize the value of the "common people" and to make the older people satisfied with their present rural status. There is a balance in the system since a sufficient number of older people stay to operate the farms and the services in Homestead center, and a number of young people leave for careers elsewhere. Thus Homestead is kept going as a community, and there is not too much population pressure on the land at any given time.

CONCLUSIONS

In the foregoing chapter we have seen that there has been a *selection* of the "future" rather than the "present" or the "past" as the significant time dimension. The *regulatory* functions of this value-choice are clearly observed in Homestead today in the persisting cultural prescriptions which dictate the desirability of living for the future, being progressive and optimistic, and continuing to strive for success. So deeply have these values been internalized that the Homesteaders say: "We believe in the future, just like everybody does anywhere else." It is inconceivable to them that other cultural groups may live more for the present or past. The orientation also plays a functional role in the *ranking of goals* for future action, for greater rewards and prestige are attainable by those who work hard for progress and success than by those who are unprogressive, pessimistic, or satisfied with the present state of affairs.

It is clear that this future-time orientation was a critical factor in the original settlement and early development of the community. Without this orientation it is doubtful that the Homesteaders would have been motivated to shift from tenant farming and farm labor jobs on the Plains of Texas to frontier dry-farming in western New Mexico. Because they were strongly oriented toward the future, they withstood the hardships of pioneer life and built a community based on dry-land farming techniques in an unsuitable region, while government officials tried to discourage their long-range hopes for success.

It is evident that the future-time orientation also plays an important part in the continuing existence of the community. For example, if the bean crop fails from drought or frost, the Homesteaders can always look forward with the hope that next year will be better. In short, attention is continually shifted from the grim realities of the present to the elusive but persisting dreams and expectations for the future.

Summer Rodeos Are Favorite Pastimes

Young Homesteaders Begin to Work in the Bean Fields at an Early Age

5

Working and Loafing

The dictates of any economic system obviously require a certain expenditure of time for sheer survival, and a certain amount of sleep is needed for continuing physiological well-being, but what is customarily done during the remaining time is an important aspect of the value-orientations of a given culture.

Frontier life in the American West has traditionally been regarded as a type of life which places great emphasis upon diligence and hard work. The pioneer is thought of as a person who by hard work transformed the wilderness into civilized communities. While such a characterization may have held for most frontier communities, the value patterns of twentieth-century Homestead, settled as part of the Southern migration across the United States, are distinctly different in this area of life. The Homesteader believes in hard work, but unlike the neighboring Mormons, he does not consider it necessary or desirable to work hard all the time. He appears to lack the Protestant ethic which prevailed in the early settlements of Puritan New England [1] and which still prevails in the industrious American middle class, whose members are not happy unless they are constantly busy and whose vacations are almost as hectic as their working days.[2] Neither does he spend his time as leisurely as do the Spanish-Americans or Navahos.

The value-orientation was expressed succinctly by one of the more articulate residents of Homestead, as follows:

> I guess the people in Homestead expect to work hard in the working season and then loaf hard in the loafing season. In the farming season, they get up early and work twelve or fourteen hours a day and enjoy it. But when that work gets done, it's hard to do more than cut wood and do the chores.

The people add that one thing they like about living in Homestead is the fact that "you can take it easy . . . you don't have to work the year around." They say, "we do more livin' here in two weeks than those city people do in a whole year."

In other words, the value-orientation appears to be a patterned combination of working and loafing or, in Florence Kluckhohn's terms, a stress upon "doing" balanced by an equal emphasis upon "being." [3] The role of this value-orientation in the life of Homestead is most clearly understood in the wider context of the daily and annual round of life.

DAILY ROUTINE

"Early to bed and early to rise" is the prevailing pattern in Homestead. Families usually get up about sunrise. The father or one of the older sons builds the fire and does the chores (milking, feeding and watering livestock, bringing in wood and water) while the mother, perhaps with the help of an older daughter, prepares breakfast. If there is time before the school bus arrives, the older children may help with the breakfast dishes. The mother may then devote her morning to cleaning the house, laundering, sewing, ironing, canning, or other tasks while the father works in the fields (during the farming season) or goes "up town" to loaf (during the winter "off season"). Dinner is eaten at noon, and the "evening" [4] may be devoted to the continuance of the morning's activities, or the mother may go to the village center to shop, visit, and pick up the mail. At sundown the chores are repeated; supper is cooked and eaten; and the family is in bed soon after dark.

The daily routine varies, of course, with such events as shopping trips to the city, dances and movies in the schoolhouse, visits of neighbors and relatives, etc.

ANNUAL ROUND OF LIFE

After a winter period of relative inactivity, farming operations begin in late February or early March with the listing of the fields. This period is one of hard and consistent work for two or three weeks because the listing cannot be started until the ground thaws and dries after the winter snowfall, and must be completed before the March windstorms become intense. If the process is not completed with dispatch, a much greater loss of topsoil may result. Often the listing operations will be interrupted by the arrival of early windstorms, and on these days the Homesteaders may gather in the

WORKING AND LOAFING

village to loaf and visit or in one of the homes to play poker. During late March and April there is a loafing period, with some work done on fences, corrals, and perhaps house-building or repairs. The Junior Prom, the most important dance of the spring, occurs during this time.

By early May the farmers are looking forward to planting time and begin to repair their planters and tractors for the next farming operation. Since corn is not so sensitive to frost as pinto beans, it can be planted in mid May. Then follows another period of inactivity until a decision is made about when to plant the bean crop. Although farmers customarily do a great deal of loafing during this period, they are tense and anxious, and the leisure time is not relaxed as it is in the winter. Indeed, much of the loafing time is spent in serious discussion and in watching the weather in order to make the best possible decision about the time to plant the beans. The Homesteader must weigh several alternatives carefully, for if he plants early, the beans may be destroyed by a late spring frost or later by drought before the summer rains arrive. If he plants late, all the winter moisture may have left the ground and the soil will be too dry, or there may not be time for the crop to mature (a minimum of ninety days) before the September frost arrives. The time selected usually falls between the 25th of May and the 10th of June. Once the decision has been made, there is again a period of intense activity until all the planting is finished. Since summer vacations for school children always begin by the third week in May, the older children are available to help with the planting operations.

The summer season is characterized by alternating periods of hard work and leisure. Immediately after planting time there is a period of anxious waiting during which a late spring frost may destroy the young bean plants and necessitate replanting. Or, if a heavy rain occurs soon after planting, it may be necessary to "scratch" the soil in order to break the crust and allow the bean plants to come up.[5] When the beans are up about four inches, the first cultivation with the knife-sled takes place. Then there is the "long wait" through the usually hot, clear, dry weeks of June and into July when the summer rains are expected to arrive. If the rains do not arrive until late August, the crop is often killed by drought or produces so little that the farmer does not even "get his seed back." If the rains do arrive, there are usually two more cultivations with a duck-foot cultivator, for by this time the bean plants are too tall for the knife-sled to be used.

The periods between these farming operations allow ample time for loafing in the village, for playing baseball, for attending Fourth-of-July and other summer rodeos and dances, and for going fishing in nearby lakes or more distant mountain streams. Homestead sponsors a two-day rodeo in the village during the summer, and each of the nearby communities also has rodeos and accompanying dances. There are also periods of leisure on rainy days when it is impossible to work in the fields, and the Homesteaders can loaf, visit, or play poker.

If the summer rains have been adequate for a crop, the early days of September usher in another tense and busy season, for it is during this period that an entire season's work and investment could be destroyed by a single heavy frost. If there has been insufficient summer precipitation or if frost destroys the beans, the farmer may start planting a crop of winter wheat (or sometimes rye) on the fields that failed to produce beans. When the frost holds off until after the 15th of September, however, the crop usually matures well, and harvesting operations begin soon thereafter. Almost a month of bean cutting, piling, and combining follows, and school is officially dismissed for at least a week to permit the children (and teachers who have farms) to help with the harvest. It is not until the crop is completely harvested and the beans stored away in the warehouse that the farmers can again afford to relax. "After harvest," they say, "we just coast along until spring."

With harvest over, the life of the village revolves around the fall hunting season which legally is from November 10 to 20. Although the Homesteaders have been hunting off and on during the leisure periods of the summer, the "official" season is a time when every able-bodied male is expected to go hunting for deer and turkey. Many of the wives also participate at this time. Weeks in advance of the season, the "talk" in the loafing groups turns to hunting stories and plans for the trips. By November 9th the hunting parties are already out in the mountains south of Homestead. Those who are successful in the opening days of the season will return early, but the other parties may stay out all ten days. After the season, accounts of the hunting trips again provide the loafing groups with a conversation topic for several weeks.

WORKING AND LOAFING

By late November the Homestead area begins to feel the force of the winter cold, and the Homesteaders must put in some time hauling wood for heating their homes. During the fall they are more likely to haul wood as it is needed, but at this time a more concerted effort is made to lay in a larger supply.

The balance of the winter, from late November through early February, is generally recognized as the "real loafing season" and the "poker season." During this three-month period there is little to do (from the Homesteader's point of view) beyond the morning and evening chores, hauling an occasional load of wood, or making needed repairs on fences, houses, or corrals. Thanksgiving is ceremonialized by a school program and family dinners. The latter include the nuclear family and sometimes members of the extended family, but the custom of large family dinners which include all available relatives (and perhaps friends) is conspicuously lacking in Homestead. At Christmas time there is another school program, a church program, and a dance, but again no large family gatherings. New Year's Eve is also celebrated by a dance, but aside from these occasional larger social events, the winter loafing season passes quietly and leisurely as the Homesteaders gather daily in the village and wait for the spring farming season.

WORKING PATTERNS

The Homesteaders' orientation to work falls between that of the Spanish-Americans and that of the Mormons. The former group is regarded by the Homesteaders as being "shiftless and lazy" and "not gettin' anything done"; the latter is recognized as having much more diligent work habits. When the Homesteaders claim to be more industrious than their Spanish-American neighbors, they do not mean that they work all the time but that they buckle down to disciplined and sustained work during those times of the year when there are crops to plant or harvest, or when there are houses, fences, or corrals to be built. In other words, they do not always let things "drift along" from day to day nor "put off work until tomorrow," for a farmer who did would be branded as "lazy." In this respect they share the general value-stress in America upon activity and hard work.[6]

The Homesteaders' conception of work is definitely that it consists of manual labor — e.g., farming, repairing machinery, and building houses. Less value is attached to such work as managing a store or engaging in trading activities. The activities of research workers, who merely talk to people and write books, do not resemble work at all. Nor is there any comprehension or appreciation of spending leisure time in intellectual or artistic pursuits. When informants were asked what they would do if they should suddenly and unexpectedly inherit $100,000 from a rich relative, they all replied that they would go on farming and ranching. Some thought they would take time off to go fishing and hunting, but no one considered complete leisure a possible way of life.

On the other hand, the few persons in the community who are always busy are singled out for special comment. Jokes are told about one of the older women of Homestead; for example, it is said:

> She has slowed down a little bit in recent years; you can now catch a glimpse of her when she turns the corner to come back down the row while she's hoeing in the summertime . . . One year she thought the crop was ruined because her husband didn't get to work until sunrise and quit at sunset!

LOAFING PATTERNS

Although a few members of the community are characterized by inner drives for constant activity, most Homesteaders, and especially the men, are quite capable of loafing. The most highly respected persons in Homestead can be found quietly talking and "whittling away" the time in one of the loafing groups in the village.[7] These groups gather daily at the two repair shops, at the two stores, and at the bar.

The groups at the shops are composed exclusively of men, who join the group after they have finished their morning chores at home, or stay to talk while repairs are being made on a piece of farm machinery or an automobile. The personnel shifts from time to time during the day and the size of the group may vary from two to twelve individuals. The managers of the shops nearly always join in the general conversation and sometimes interrupt their work to tell a story. The school-bus drivers who have deposited the children at the school and "have a day to put in" before taking the children home are also fairly steady participants in the loafing groups. The managers of the shops obligingly provide a few boxes and empty carbide or oil cans to serve as seats for the loafers. When all the seats are taken, the others squat on their heels or lean against

WORKING AND LOAFING

the wall. In winter the group gathers around the stove and in summer often outside in front of the shops.

These all-male groups engage in talk, rumor, and gossip about community affairs; many of them whittle and chew tobacco or smoke; sometimes they play dominoes, "moon" being the favorite game. The talk is of weather, crops, and livestock; of hunting and fishing; often of politics (but almost never about religion); of recent trips made to the city; and of the latest "scandal" in the community — the men gossip almost as much as the women in Homestead. A large proportion of the time is spent telling "tall tales," and several of the older men enjoy community-wide reputations as good storytellers.

In the stores, benches are provided for the loafers and are occupied by both men and women who either come to make purchases and stay to visit or just drop by to get the latest word on community affairs. The loafing in the stores is more frequently by family groups. To an urban observer the time spent in these loafing groups seems almost endless. The writer has many times attempted to sit through one of these loafing bull sessions, but found it almost impossible to do so.

The bar provides still a different atmosphere for loafing, since it is perhaps the most controversial place of business in the community. Homestead has never been without a supply of liquor. When the village was first settled, prohibition was still in force, and the liquor supply came from "corn likker stills" hidden in the woods. After the repeal, a bar was established for a few years in the center of town. It closed, but after World War II another bar was opened on the north edge of the town center. It is owned by a Spanish-American who tends the bar and runs a poker game whenever anyone wants to play.

The women of Homestead are almost unanimously opposed to having a bar in the community. The men are divided in their opinions; most of them drink at least to the extent of having a beer now and then but some, particularly those who are Baptists, are strongly opposed to the bar. There is a strong taboo against women going to the bar "except to drag their husbands out and take them home," and the taboo is strictly observed except by one or two women in the community. Thus, the bar becomes almost a men's drinking club, similar to the old-fashioned saloon. It does a little business every day when a small group of men gather to drink beer, talk, and perhaps play poker. It does a booming business when there is a dance in the schoolhouse "and the men beat a path from the schoolhouse to the bar and back."

The women's pattern of work and leisure is somewhat different from the men's. In the first place, there is not the same cycle of busy farming season followed by winter loafing for the women, whose tasks of housekeeping, cooking, and child care go on the year around. Although the Homestead women are not "pressed for time" as are the urban middle-class women who attempt at the same time to keep house and engage in many club or community activities, they nevertheless do not spend entire days loafing as do the men. They are more likely to spend only part of the day, when they have finished their usual work, visiting in the stores or with other women on neighboring farms. Furthermore, the women are expected to act as corrective agents for men whose loafing becomes unreasonably frequent and to prod the men into more useful activities. For example, one afternoon when a woman's husband and three other men were supposed to be installing the plumbing in her new house, the woman caught them all playing poker. She "raised hell and made them get back to work," and after that went each day to the site to see that there would be no more loafing or poker playing until the house was completed.

Although most of the Homesteader's leisure time is devoted to visiting, gossiping, and loafing in informal groups around town, there are a number of organized recreational activities which play an important role in the functioning of the community. The most important in terms of community support and numbers of people involved is the baseball club. Homestead organized a team in the second year of the town's existence and started playing baseball against teams in the neighboring communities. The Homestead team's record has been outstanding; year after year they win over teams from the larger towns and take great pride in this fact. The games bring out almost every family in the community, thus uniting many of the factions in town. Although there is continual bickering and feuding as to who should play on the team, the team functions competently when the game begins. Through this activity the competitive attitudes which govern social action within the community are re-channelled into competition against other communities, and

WORKING AND LOAFING

the fact that Homestead wins a majority of the games has high symbolic value as a means of showing these older and larger communities that Homestead is a town to be "reckoned with."

The dances in the schoolhouse provide another important focal point for community activity, for although they are objected to by the Baptists and many others, they are nevertheless attended by a substantial majority of the families in Homestead. The dances take place in connection with holidays: Christmas, New Year's, Fourth of July, and Thanksgiving; there are one or more dances during the summer rodeo, one on President Roosevelt's birthday, and a Junior Prom which is the most important school-sponsored dance during the year. It is also possible for an individual family to organize a dance at other times (usually to celebrate the arrival of visiting kinsmen from Texas or California).

The ingredients for a good dance include (a) a "big crowd," and posters are put up in the stores and post office in Homestead and other towns to advertise the dance; (b) "good music," which is provided by local "fiddlers" and guitar players who always play "Western" or "Hillbilly" tunes; and (c) "salty dog" for the men. Women do not drink, but the men bring "salty dog," which is a mixture of bourbon, grapefruit juice, and salt, and keep the bottles outside in their cars. The dances begin about 9:00 p.m. and last until well after midnight. The women sit on chairs or benches along the edge of the auditorium in the schoolhouse; the men stand around the door or out in the hall, and when the music begins they seek out partners, dance the two parts to a "set," and return the "ladies" to their seats. Between dances the men dash outside to drink "salty dog" while the women sit primly inside to wait for the next dance. Dancing mainly takes the form of a fast two-step, but from time to time there is a schottisch, a varsouviana, or one or two "Paul Joneses" (a form of the quadrille) during the course of the evening.

While dancing is opposed by certain religious segments of the community, "42" (a domino game) is played by almost everyone. It is characteristically a "Texan" game in this area of New Mexico and is unknown to the Anglos, Mormons, and Spanish-Americans. Back in Texas there were large "domino parlors" where the men played "42." In Homestead the game is played at "42" parties which are given to celebrate a birthday, or it is played in an evening or on a Sunday when two or more families get together for a visit. Both men and women play, whereas the other domino game, "moon," is usually played only by the men while they loaf at the shops.

Poker is regarded as "sinful" by many families in Homestead and the game is not as widely played in the community as "42." For the most part, the poker playing takes place at the bar, where the bartender runs a no-limit game, or at the homes of the six or eight men (ages twenty to forty) who gather to play "penny ante" during periods of inclement weather.

SOME COMPARATIVE NOTES

It will be noted that the value-orientations we have discussed previously may be said to accord with what has been described as the "dominant orientation profile" of American culture.[8] The "individualism," "mastery over nature," and stress upon the "future" are all conspicuous features of the generalized American value system. In the case of the value-orientation discussed in this chapter, the Homesteaders manifest a value-complex that differs significantly from the dominant profile with its emphasis upon "doing," "achieving," "activity," and consistent "hard work." The Homesteaders have instead been described as having a balanced combination of "doing-being," or "working-loafing" — a value position which rewards those who can work hard *and* loaf but which subjects those who engage in constant work and activity to criticism and ridicule.

It is illuminating to compare this Homesteader value-orientation with the Mormon patterns as observed in the neighboring village of Rimrock. In the Mormon life-way there is no place for loafing, because if there is no work to do for one's self, there is always work to do for the church. Occasionally two or three people may be seen visiting briefly in the store or in the repair shop, but there are no long loafing sessions. The store and shop have never had benches or seats for loafers, except at a time when the general store was managed by a Texan. He placed chairs around the stove with the expectation that they would be used by his Mormon customers. Instead, the chairs were used by the Navahos who came to shop in the store and usually stayed all day. When this same store was purchased by a Mormon family, the chairs were taken out, and the Navahos now have to lean against the counter, sit on the floor, or spend the day outside in front of the store.

This "activism" of the Mormons has been well described by

WORKING AND LOAFING

O'Dea who shows that it has become one of the central values of Mormon culture and permeates the economic, social, and religious life of the group.[9] When Mormon families are not busy in their farming, ranching, or housekeeping activities, they are attending church-sponsored meetings which occur several times a week. Even the apparently recreational pursuit of dancing is imbued with "activistic" values in Mormon culture, where the ability to dance and the obligation to participate in the community dances are strongly emphasized. Whereas the person who is constantly busy and not happy unless he is working warrants special comment and criticism in Homestead, the loafer in Rimrock is singled out for condemnation.

It is the writer's contention that the working-loafing pattern in Homestead is not a direct response to the annual requirements of bean farming, for there is no situational reason why the rainy days during the farming season or the long winter season could not be devoted to more constructive activity. If Homestead were populated by a group of German or Norwegian farmers, for example, it is clear that their drive for constant work and activity would not permit them such endless periods of loafing. In rural "Jonesville," where the writer did previous field work, the midwestern farmers enjoyed a leisurely shopping trip to town on Saturday, but the rest of the time (even during the winter season) were generally so busy that it was difficult to arrange interviews with them.[10]

It is also evident that the Homesteaders' cyclical working-loafing values are characteristic of the subcultural group of which the Homesteaders are a small part. The same activity pattern is found in the hill country of the Southern states, on the Great Plains, and elsewhere in the Southwest. In his study of Cotton Center, Texas, Bailey found that there were always many street-corner bull sessions. "People could be located according to the bench sat upon, or the store-front leaned upon."[11] Of the farmer in the wheat-producing area of Kansas, Bell comments:

> In the midst of a busy season, he gets up early and goes to bed late, and he may work more than 20 hours a day. If there is not much to do, he may be away all day and do what little work he must when it is most convenient. The Haskell County farmer is sometimes called lazy. But he works hard when there is work to be done.[12]

In his study of *Plainville, U.S.A.*, James West has described at length the "loafing groups" which characterize the life of this village in the hill country of the Ozarks.[13]

The Homesteaders' value-emphasis on loafing which they brought with them when they migrated was reinforced by life in New Mexico, because the bean crop did not require constant work during many months of the year as did the cotton crop in Texas; thus it was possible for them to give full expression to this value-orientation.

CONCLUSIONS

In the foregoing description of the doing-being orientation of the Homesteaders, we have seen that there has been a *selection* of the pattern of working hard when there is vital work to do and of "loafing hard" the rest of the time. They have not selected the possible alternative patterns of consistent work and activity that characterize the Mormon group or the more relaxed pace of living that characterizes Spanish-American groups. The selection has not been merely a response to the situational requirements of bean farming in this New Mexican setting but represents an older style of life which appears to be an integral feature of the life-ways of the particular subcultural group found in Homestead. The pattern has a widespread geographical distribution, extending from the hill country of the South through the Great Plains and on to the Southwest.

The *regulatory* function of this value-orientation is plainly manifested in Homestead today by the definitions of appropriate behavior for individual farmers as described above. There is also consistent criticism and ridicule both of the few Homesteaders who neglect their important farming work and of those few who are constantly busy and feel compelled to work all the time.

The orientation is also manifested in the *goal-discriminations* in the cultural system which constantly remind the Homesteaders, as they examine their alternatives for future action, that the leisure they enjoy in Homestead would not be forthcoming in wage-labor jobs in the nearby cities, and that despite the crop failures Homestead is, after all, a desirable place in which to live.

It is evident that this balanced working-loafing orientation has both positive and negative effects upon the continuing existence of Homestead. The effect is unequivocally positive in so far as the opportunity for leisure influences families to remain in Homestead rather than seek more economically advantageous wage-labor jobs

WORKING AND LOAFING

outside. Next to independence ("being your own boss"), the most frequently mentioned reason for staying in the community is that "we can get by without working too hard . . . We work, but we also have plenty of time to loaf, play ball, and dance."

The negative effects are equally clear. There is no objective reason why the time devoted to loafing in the village could not be devoted to community improvements, such as finding solutions to Homestead's economic and social problems and working on new school buildings. After the Homesteaders refused to volunteer their labor for making adobes for the high-school gymnasium on the grounds that they would be busy on their "places," enough time was spent loafing in the village in the next two months to have completed the job easily.

While the casual observer might perceive an inconsistency between the value placed upon progress, the future, and personal achievement and, on the other hand, the pattern of loafing, the Homesteader does not feel he is being inconsistent. He feels that if he works while there is necessary work to do, then he may relax and enjoy his hours of loafing — which give him time to brag about his personal achievements, about the progress he feels the community has made, and about the prospects for the future.

6

Group Superiority and Inferiority

The outstanding feature of the cultural environment into which the Homesteaders moved in the 1930's was its diversity. The Homesteaders came from an environment uniform by comparison; in the small towns of western Texas and Oklahoma there were differences in religious affiliation, but only between Protestant sects; differences in social class were apparent, but there were certain common commitments to the American way of life; and although differing in race, the Negroes nevertheless spoke English and were culturally similar to the whites. In western New Mexico the Homesteaders found themselves neighbors of Pueblo and Navaho Indians who had completely different religions, different values, and who spoke alien languages; of Spanish-Americans who were Catholic in religion, maintained "strange" customs, and spoke Spanish by preference; and of Mormons who looked like the Homesteaders, but whose religion and customs were not at all the same.

The oldest settlers, the Pueblos, were found in their present location by Coronado in 1540, and it is now clear that they are cultural descendants from the earlier Pueblo occupation of this part of the Southwest. The earlier Pueblo Indians already raised corn and lived partly by agriculture and partly by hunting. Their archaeological sites are found throughout the Homestead region which they occupied from approximately 500 A.D. to 1350 A.D. The sites range in size from one or two rooms up to large villages of 500 or more rooms. Early in the period of occupation the sites were very small and scattered, but later (especially after 1000 A.D.) the Pueblos began to congregate in larger villages.[1] By the time of the arrival of the Spanish conquistadores, these sites had all been abandoned and the Pueblo peoples were concentrated in the present Pueblo villages. However, the old trail from Pueblo to Salt Lake passed through the center of the region that is now Homestead, and there has been continued Pueblo travel through the area.

GROUP SUPERIORITY AND INFERIORITY 123

Pueblo now has a population of approximately 3000, located in one central and four small farming villages. There are irrigation systems in each of the settlements, and the Pueblos operate large herds of sheep and cattle and do a flourishing business in silversmithing.

The Navahos lived in the region for some time before they were taken to Fort Sumner in 1864, and after their release from captivity in 1868 a few related families returned to this region instead of settling on the Reservation. There are now approximately 650 Navahos occupying the area between Homestead and Rimrock. The subsistence pattern can be described as a combination of sheep husbandry and dry-farming agriculture but with an increasing emphasis upon wagework on the railroads and seasonal agricultural labor in various parts of the intermontane West.

The Spanish conquistadores were first in the area with the Coronado expedition of 1540 and several later Spanish expeditions passed through the region. The immediate ancestors of the present Spanish-American inhabitants, however, did not settle in the region until the 1870's and 1880's, when they moved west from the settlement of Cubero and established themselves as ranchers in three small communities near Homestead, where they number approximately 100.

The Mormons arrived from Arizona and Utah as missionaries to the Indians during the 1870's and 1880's and founded a missionary outpost in the Rimrock valley. They built an irrigation reservoir and a community based upon irrigation farming. The present population of approximately 250 depends upon farming in the valley but is being increasingly drawn into dry-land ranching to the south and east of the village and, to a lesser extent, into wagework in the nearby urban centers as the population expands and exceeds the capacity of the original land base.[2]

In their relationships to these other cultural groups in the Southwest and to the larger outside world of cities and "city folks," the Homesteaders manifest a complex combination of attitudes of superiority and inferiority. Although the Homesteaders believe in the cherished American value of "equality," they apply it strictly to relationships among "white folks" and not to their "Mexican" and Indian neighbors whom they have tended to equate with "niggers" and to regard as inferior in race, culture, and standards of living.

This Texan Homesteader attitude contrasts sharply with the orientation of the Southwestern Spanish-American and Indian people, who manifest a much greater tolerance and respect for different ways of life.

On the other hand, the Homesteaders manifest very evident feelings of inferiority and uneasiness in their relationships with the metropolitan world. They are particularly sensitive to the fact that their town is a "backwoods community" and that their customs and dialect are those of Southern "country hicks." As they face the larger American society of which they feel they are a part, but a "backwoods" part, they are never quite certain that they are being taken seriously by those in more central positions of power, influence, and respectability. In this direction, the self-reliant and independent Homesteader manifests strong feelings of inadequacy and inferiority and a deep desire to validate his status and increase his security in contacts with the general American society and the larger stream of events.

To a certain extent these orientations of inferiority are also evident in contacts with the "big ranchers" and with the Mormon community of Rimrock, although there is much more ambivalence expressed by the Homesteaders with regard to these groups.

It is the purpose of this chapter to trace the effect of the complex group superiority and inferiority orientations upon the relationships which the Homesteaders have developed with their neighbors of varying cultures and with the larger urban world.

EQUALITY

Although differences in social status exist among the Homesteaders, especially between the "Tobacco Road" fringe and the "acceptable" families in the community,[3] the doctrine of "equality" is firmly held as far as relationships among people within the community are concerned. Everyone is welcome at the dances held in the schoolhouse, where each man is expected to dance with most, if not all, of the women present in his own age-grade. Any family visiting the house of another family at mealtime would be invited to eat. Marriage may take place between any persons of different families in Homestead, and it is believed that each nuclear family within the community should have an equal voice with every other in the management of community affairs.

GROUP SUPERIORITY AND INFERIORITY 125

GROUP SUPERIORITY ORIENTATIONS

When the writer began his interviewing on the subject of "equality," he was told by one of the older residents of Homestead: "Well, of course, we're a bunch of Texans and we don't go for racial equality. You know that." It soon became apparent that the caste relationships which the Homesteaders maintained with the Negroes in the South had been generalized to include the "dark-skinned" Spanish-Americans and Indians in New Mexico. This attitude still persists in Homestead in the tendency to regard both Spanish-Americans and Indians as "black," "shiftless," and "uncivilized," and in the persisting feelings against inter-dining, inter-dancing, and intermarriage. In 1950 when I asked one of the older women (who had lived in Homestead for two decades) what she thought about "equality," she replied (with affect):

> Well, you're talkin' to one who don't believe niggers are equal to whites. I don't believe for a minute that we should be sittin' down with them, eatin' with them, or goin' to school with niggers.

When I pointed out that there were no Negroes living in Homestead or in the immediate area, she continued without a pause:

> Well, it's hard here, too, with the white kids having to go to school with them Mexicans. Some are awful dirty and have different diseases.

Yet, certain small changes have taken place in the attitudes toward Spanish-Americans and Indians as the Texan Homesteaders have experienced twenty years of inter-cultural contact.

RELATIONSHIPS WITH SPANISH-AMERICANS

Before the Homesteaders migrated to New Mexico their only contacts with Mexicans were with the migratory laborers who came in small numbers to the Panhandle each year to work in the cotton harvest.[4] These Mexican laborers occupied a depressed position within the social and economic structure; as one of the Homesteaders expressed it: "Down in Texas, Mexicans are considered as niggers, and niggers are pretty low." Although the Texas Panhandle was not part of the original area of Mexican settlement and actual political control, this stereotype of the Mexicans was reinforced in the minds of the west Texas families through their school lessons in Texan history and folk-lore about the battle of the Alamo, in which the Mexicans were pictured as the treacherous villains and the Texans as the courageous and righteous heroes. In addition, west Texas is part of the Southern "Bible Belt," in which feelings against Catholicism have always been strong.

It is not surprising that the Homesteaders, with this background, were distressed when they settled in western New Mexico and discovered that almost all of the county and many of the state government officials were "Mexicans"; that they would be living next to the "Mexican" village of Tapala where the people were Catholic and spoke Spanish in preference to English; and that the post office and high school (staffed by "Mexican" teachers) were located in Tapala. From the beginning the Homesteaders manifested a distrust and fear of these people. One of the older Homesteaders, on his first trip to Tapala, was invited to eat in the house of the *patron* of the village. He was afraid to go inside the house but accepted a plate of food and remained outside the house while he ate. One of the younger Homesteaders stated: "It was almost like going to nigger town when we had to go to Tapala for the mail. We kids used to carry sawed-off shotguns with us in case there was any trouble."

The Spanish-Americans, who had developed a modus vivendi with the older Anglo-American ranchers and settlers in the region in which there were relatively tolerant and respectful relationships between the two groups, were quick to sense that these "Tejanos" manifested hate, prejudice, and superiority feelings toward "Mexicans." They also soon discovered that the "Tejanos" posed a serious economic threat to their village when the Homesteaders began to acquire title to, and fence in, land which the Spanish-Americans had been grazing as open range land for almost fifty years. Although they first made overtures of friendship to the "Tejanos," many of whom they pitied because of their utter poverty, they eventually withdrew such gestures after rebuffs by the Texans and later joined the Anglo ranchers in predicting that Homestead would not exist for more than a few years and that it would be impossible to farm in this arid region.

In the fall of 1934, an outbreak of the repressed hostility between the two communities occurred when a Spanish-American teacher (and son of one of the most respected families in Tapala) was supplied by the county for the newly constructed grade school in

GROUP SUPERIORITY AND INFERIORITY 127

Homestead. This move met with violent objections from the Homesteaders; during the first few weeks of school, windows were broken at night and signs appeared upon the door of the school which read: "We Don't Want Any Chile Pickers for Teachers!" and were signed, "KKK" (Ku Klux Klan). Finally, one night the schoolhouse burned down, and both Spanish-Americans and some of the Homesteaders now state that it was set on fire by one of the more rabid Mexican-hating Homesteaders, but the man was never formally confronted or punished for the crime. The Homesteaders then set up their own school system, employing one of their own group to teach the school, and charging five dollars a month for each child rather than relying on the county, which would have provided funds but might also have provided another Spanish-American teacher.

These events, of course, increased tension on both sides so that the Spanish-Americans for a time avoided going to Homestead, and the Homesteaders were rather roughly treated if they attended dances in Tapala. On one occasion when a fight threatened, a group of Homestead men had to leave the dance early, and as they drove away rocks were thrown at their truck. Yet during this time the older Homestead children were attending high school in Tapala, and families still had to go to Tapala to pick up their mail. This inevitably brought the two groups into contact, and a few friendships began to develop between individual members of the two villages. Most of the Homesteaders, however, continued to object to sending their children to the high school in Tapala and began to talk openly of attempting to take both the school and the post office away from Tapala. They also envisioned roads which would connect their own village with urban centers and leave Tapala off on a side road and looked forward eagerly to the day when their numbers would bring them political control over the Tapala district.[5]

Eventually most of these things came about. By 1936 Homestead had its own post office, although a small post office is still maintained in Tapala. In 1938 the high school was shifted to Homestead, and a few years later the grade school which had been maintained in Tapala was closed; the Spanish-American children now must come to Homestead to school on the bus. In 1950 a new road was built directly into Homestead from Gallup, leaving Tapala off on a side road. The population decline of Tapala, which began when earlier Anglo-Americans started to acquire control of the range land, was speeded up by Homesteaders acquiring additional lands, and by 1951 the Spanish-American village had less than fifty residents.

Gradually, however, a few closer ties have been established between the two villages. Despite the objections of friends and relatives, two of the Texan men married daughters of the former Spanish-American *patron* of Tapala, and these two families still reside in Homestead. A young Spanish-American girl from San Fidel was sent to teach the Homestead grade school and was well liked by the Homesteaders. One of the Spanish-Americans set up a bar in Homestead which is still well patronized by the drinking part of the population. Members of the two communities participated in one another's dances, and in the baseball games and rodeos in Homestead.

Thus it appeared for a time that better relationships between the two communities were developing, but in 1947 open conflict broke out in a form more serious than before. During a dance in Homestead a fight started when one of the young Homesteaders pushed a Spanish-American at the door and said, "Out of my way, you dirty Mexican!" Versions of the fight vary depending upon which group tells the story. According to the Spanish-Americans, there was a fair fight between the two men but when the Spanish-American won, the young Texan staggered back into the dance hall and reported that he had been beaten up by "a whole gang of Mexicans." The Texans then came out in force and proceeded to beat up all the Spanish-Americans they could lay their hands on. The Texan version is that the "Mexicans" were resentful of the fact that a Texan couple had won the jitterbug contest, and that several "Mexicans" had "jumped on" the Texan and beaten him in an unfair fight.

At any rate, it is clear that almost all the Spanish-Americans and many of the Texan men present were involved in the fighting before the night was over. While the son of the former *patron* of Tapala had tried to stop the fighting, the important leaders of Homestead who were present that night either joined in or indicated that they would if it became necessary in order to win the fight. One of the most respected Homestead leaders is reported to have taken off his coat and said, "If any damned Mexican wants to fight, just let him step up."

The fighting, which is now called the "Spanish-American War" by the Homesteaders, continued off and on for several days as the young Texan who was beaten sought revenge on all of the young Spanish-Americans who had taken part in the fight. He challenged to a fight each of them who came to Homestead. He fought the school-bus driver from Tapala, as well as a man who had merely driven over to Homestead to haul a barrel of water from the community well. During the same period there was talk in the Veterans Administration farming class that the Homesteaders "ought to run the Mexican GI's off and not let them come to class any more." The teacher effectively put a stop to this plan by reminding the Texan members of the class that the Spanish-Americans would write to Senator Dennis Chavez about it, thereby perhaps putting an end to the GI class and the monthly subsistence checks. The fight also had repercussions in the high school, where the Texan students told the younger brother of one of the Spanish-Americans who had been in the fight that if he continued coming to school, they would beat him up. In response to the threats, the Spanish-American went to Gallup to continue high school.

Although the open fighting stopped after a week or so and many Homestead families who were not present at the fight regretted the incident, the "Spanish-American War," like the school conflict fifteen years before, was another tension-producing event affecting the relationships between the communities. It seems unlikely that these events will soon be forgotten by either group.

Despite this troubled history, there has been a modification of the strictly higher and lower caste relationship between the two groups. After the Homesteaders overcame their early fear of going to Tapala ("like going to nigger town"), the two communities began to attend each other's dances and eventually there were cases of inter-dining and intermarriage which certainly would not have occurred with Negroes in Texas. Homesteaders also now recognize certain high-status families in Tapala who, they feel, deserve more respectful treatment and are spoken of as the "whiter type of Mexican." Furthermore, the Homesteaders admit that the Spanish-Americans have certain cultural techniques worth adopting, such as building houses with adobe bricks, now a general practice in Homestead.

On the other hand, there is ample evidence that the "Alamo" is still remembered, that the Homesteaders still feel themselves to be superior in race and culture, and that even most of the younger generation of Texans who were born in Homestead maintain these attitudes. In 1952 when a fair-skinned Spanish-American girl was sent to teach in Homestead, the Texan students were amazed to find that she was not black. Intermarriage is still discouraged and the children of the mixed marriages are spoken of as "half-breeds." The actual cases of inter-dancing during the course of an evening are few, and inter-dining is still the exception rather than the rule.

It is also plain that the Texans, with their attitudes of superiority, have made little or no effort to learn about Spanish-American culture. Only one or two Homesteaders speak more than a few words of greeting in Spanish. There continues to be abysmal ignorance of the nature of Catholicism, and as of 1952 there were no Homesteaders (except for the two who had married Tapala women) who had even heard of San José, the patron saint of Tapala, or knew why the fiesta in Tapala was held on a certain day each year.[6]

RELATIONSHIPS WITH INDIANS

The Homesteaders' previous first-hand contact with Indians was even more limited than their experience with Spanish-Americans. With the exception of the few families who had lived in Oklahoma, the Homesteaders had never actually seen an Indian except in the movies. The settlers arrived in New Mexico with pioneer-type attitudes toward Indians; e.g., Indians were "wild and uncivilized," were dangerous if not closely watched, but also were destined to disappear as the land was settled by "superior" white people.

The Homesteaders' first contacts with Indians occurred during the fall and winter of the "big snow" (1931–1932), when the Navahos came in large numbers to gather the pinyon crop in the Homestead region. The Homesteaders were also picking pinyons and ordered the Navahos off their lands. One of the original Homesteaders reported:

The Indians didn't like the idea of the Homesteaders being here much. They had been used to gathering these pinyons. I had quite a time with one, trying to tell him I didn't want him to gather pinyons. And he said that the *patron* of Tapala had let him gather pinyons before. And then the Navahos stole feed for their horses from our crops. As a whole we weren't getting along very well that first time.

One of the Homesteader wives described her first contact with Indians as follows:

We were out ahuntin' pinyons and when we returned to our cabin, it was surrounded by Navahos, and it liked to scared me to death, because

GROUP SUPERIORITY AND INFERIORITY 131

I had never seen any before. I finally got over that feeling after a while, but up until I came out here I had never been around any other race of people other than my own race.

It was from these early contacts with Navahos, however, that the Homesteaders learned the techniques of robbing packrat nests and using screens in pinyon collecting.

Following this winter, contacts with Navahos were slight until the Homesteaders enlarged their fields and began to plant larger crops of beans. They then started employing Navahos as farm hands to hoe weeds during the summer and to help with the harvesting operations in the fall. For the most part, the Navahos camped at the edge of the fields, and if they were fed by the Homesteaders, their meals were served outside or separately from the rest of the family.

A few Navahos also began to trade at the Homestead stores, but although many of the Rimrock Navahos live closer to Homestead than to Rimrock, Navaho business is still negligible because the Homestead storekeepers have never encouraged it. They do not like the idea of having Indians in their stores which are "for white folks," and they have never learned the language or the trading habits of Navahos. The storekeeper of the largest store stated to the writer, "I don't want them lousy stinkin' fellows around."

The Navahos have always attended the Homestead rodeo during the summer, but for the most part they are only observers of the events. Active participation of the Indians is discouraged, especially by the judges, who tend to give the prizes to white participants even in cases of superior performances by Navahos. The Homesteaders also generally refuse to let the Navahos camp on their land when they come to the rodeo.

After World War II, three Navaho veterans enrolled in the "GI" farm training class in Homestead. The class was first taught by one of the more tolerant members of the community who made an effort to visit the Navaho farms and encouraged the Navaho veterans to raise chickens and use better farming methods. The Navaho veterans were relatively well treated in this class, in part no doubt because their presence added to the income of the instructor.[7] Two years later an advanced farming class was started which was taught by a less tolerant Homesteader. The Navaho veterans were excluded with the excuse that they were "too much trouble."

During the 1940's two Texan families (the Smiths and the Joneses) moved to Homestead from the Peña area, where they had lived on more isolated farms with Navahos as neighbors; they had learned to speak the Navaho language and to depend upon their Indian neighbors in a number of ways. These two families have continued to maintain close relationships with Navahos and frequently employ them on their farms; the Navahos eat with the families, ride into Homestead center in the same pickups, etc. The other Homesteaders single out these two families for special comment, generally failing to understand how they "can stand to have Indians eating at their tables and sitting in their houses." At the post office one day, one of the Homesteader wives commented with a shudder (when she observed a Navaho riding in the pickup with Mr. and Mrs. Smith), "I wouldn't sit next to that old black thing!" When a brother of Mr. Jones accepted two orphan Navaho children from a welfare agency and brought them back to Homestead in 1952, the wife of the storekeeper immediately cancelled her dairy milk order from the Joneses, saying that her family would not drink any milk from a house in which Navahos lived.

Contacts with the Pueblo Indians were initiated when the Pueblos traveled through Homestead following the ancient route to the salt lake from which salt had been gathered by Indians of the Southwest for many centuries. Traveling by wagon in the 1930's and by pickup truck since around 1940, these salt parties stopped to trade at Homestead stores and to visit briefly in the village. The older Homesteaders report that they learned many facts about the weather, especially about the times of frost and the best times for planting crops, from discussions with the Pueblos who, unlike the Navahos, spoke some English and could communicate with the Texans.

When the new road from Homestead to Gallup was completed across the Pueblo Reservation in the 1940's, the Homesteaders began to see more of the Pueblos. In particular, closer contacts were established with one of the Pueblo families which lived near a muddy stretch of the road and was frequently called upon to house and feed Homesteaders who were marooned overnight because of the mud. Members of this Pueblo family also came to Homestead frequently for trading and visiting.

Aside from these contacts stemming from the salt expeditions and from the Homesteaders' difficulties on muddy roads across the Reservation, there are few other important contacts with Pueblos.

GROUP SUPERIORITY AND INFERIORITY 133

The Homesteaders do not employ them as they do Navahos, since the Pueblos are, on the whole, in a more secure economic position than the Homestead farmers. Homesteaders seldom go to Pueblo except to engage in occasional trading (e.g., selling eggs to the traders, or turkeys and chickens to the Indians) or to accompany the Homestead high-school basketball team on the rare occasions when games are scheduled with Pueblo teams. Unlike the other cultural groups in the area, the Homesteaders have few close Pueblo "friends" whom they could visit in Pueblo.

No intermarriage has taken place with either Pueblos or Navahos nor, as far as the writer could determine, have there been cases of extramarital liaisons. Marriage with Indians is not regarded by most of the Homesteaders as even a remote possibility.

The technological items mentioned before are the extent of most Homesteaders' knowledge of Indian customs and values, and there is little interest in learning anything further. The Homesteaders have the ethnocentric attitude that because their own ways are so superior, there is no reason to learn about Indian customs. The only Homesteaders who speak any Navaho except for a few words of greeting are the Smith and Jones families. A few of the Texans have attended the Gallup Inter-Tribal Indian Ceremonials, but most of those who happen to be in Gallup during this time in August are more likely to go to the movies than to attend a Ceremonial. The only record of Homesteaders attending the famous Pueblo Shalako (again, with the exception of the Smith and Jones families) is in 1950, when two families were taken to this ceremony by one of our project research workers; and there is no record of attendance at Navaho "squaw dances" or "sings," which are performed by the neighboring Rimrock Navahos. These facts become more dramatic when one stops to think that American tourists come from great distances to witness these famous Indian ceremonials which occur only twenty to forty miles away from the Homesteaders.

As of 1952 many Homesteaders still failed to distinguish clearly between Navahos and Pueblos and were continually mixing up or combining the two tribes in their remarks during my interviews with them. Those who did distinguish the two tribes regarded the Pueblos as a "higher type of Indian," and spoke of them as being "more industrious farming people" who live in "better houses than the Navahos." The Navahos were commonly regarded as "the poorest and most ignorant people on earth." In comparing the three neighboring cultural groups, the Homesteaders regard the Spanish-Americans as "a higher class of people" than the Pueblos, and the Pueblos as "a notch above the Navahos." All three groups, however, are still considered inferior to "white folks."

The *persistence* of value-orientations of ethnocentricity and group superiority is, on the whole, more notable than any *changes* which have taken place in the two decades of first-hand contact of the Homesteaders with "Mexicans" and Indians. There is, of course, variation among the Homestead population, with eight families having developed somewhat more tolerant attitudes toward their neighboring cultural groups. Most of the variation can be traced to personality differences and to factors in the unique experiences of individuals in their inter-cultural contacts. Of the eight families, the Smiths and the Joneses (formerly from the Peña area) are outstandingly tolerant, and there appear to be important social-structural reasons for this. The Texans in Homestead have from the first formed a large enough cluster of families to support an all-Texan community with its own distinctive network of economic and social relationships and self-sustaining value system. In contrast, Texans in the Peña area are fewer in number and occupy isolated ranches in a setting in which their closest neighbors have been Navahos and Spanish-Americans. They have not formed a distinct community with a service center which takes care of the essential economic and social needs of the more isolated families. Instead they have come to depend more upon their Indian and Spanish-American neighbors in a situation that has promoted more knowledge about and respect for the ways of life of other peoples. In the words of Mr. Smith:

> The people in Homestead don't get along as well with the Spanish or Navaho people as I have, living way out — or somebody else way out — because they are bunched right there together. There's enough people to congregate together and help one another, and they don't need the help of the Spanish and Navaho people. I think that's what's caused it.

This explanation has been reinforced by Landgraf's findings that the relationships between Texans and Navahos at Peña were markedly friendly and equalitarian in nature.[8]

GROUP INFERIORITY ORIENTATIONS

Unlike the Mormon village of Rimrock, which occupies a firm and explicit position in the wider social structure of the Mormon "empire" and has the church to mediate between the local community and the metropolitan world, the Homesteader must face the general American social system unsupported by intermediate organizations. The local Mormon derives a sense of status security from the "concrete community of Zion," while the local Homesteader feels he is a peripheral member of American society and is never quite certain that he is being taken seriously by those who are in a more central position.[9] This feeling is frequently verbalized by such statements as, "We sure settled at the jumping-off place when we came to Homestead." It is manifested behaviorally in face-to-face relationships with urban businessmen, state politicians, and the project research workers, all of whom the Homesteaders fear may regard them as "hicks" from a "backwoods community." The attitude appears whenever a Homesteader finds himself in an urban setting, especially since the travel experience of the villagers is quite limited. For the most part, the "known world" consists of the South and the Southwest, from Tennessee to Oklahoma, Texas, and New Mexico, and west to Arizona or California. Until World War II veterans broadened these travel horizons, only one of the Homesteaders had been in New York City (where he went to participate on a "Meet the People" radio program), and only two had been in Chicago (one to attend a poultry show, the other to attend the 1932 World's Fair).

From these orientations of inferiority vis-à-vis the metropolitan world, the Homesteaders have developed strong motivations not only to prove to government officials from the Departments of Agriculture and the Interior that Homestead cannot be dismissed as an ill-advised experiment that failed, but also to make the community "more modern" and the ways of life of its members more sophisticated. These motivations are, of course, an aspect of the value stress upon "progress," but the "progress" is measured against those aspects of urban centers which the Homesteaders feel they lack.

Thus, it was no coincidence that a bridge club was organized for the first time in 1949 when our project research workers started living in the community. Nothing was done on the part of the research workers to initiate the bridge club; indeed, for a time the writer and his wife tried to discourage the idea. But the Homesteader wives insisted that they wanted to learn to play bridge, and the research workers' wives were asked to be instructors. During meetings of the club the Homestead women often made comments like, "Aren't we becoming ritzy?" and, "We're just like the 400 crowd." When we returned to the community in 1951, the bridge club was still meeting from time to time and had become a symbol of reassurance to the Homesteader wives that the community was becoming "up-to-date" in its social life.

Comparable responses were observed in the Homesteaders' sensitivity to their rural Southern speech habits, their clothing, table manners, houses, etc.; they felt they must apologize, and frequently asked questions as to how urban people talk, dress, eat, and furnish houses. The research workers generally followed the policy of trying to reassure the Homestead families that Southern rural life had positive values that in many respects were preferable to city ways. Despite these efforts, however, the sensitivity and the striving to become more sophisticated continued unabated and was verbalized whenever either another research worker or an urban businessman came to the community.

RELATIONSHIPS WITH RANCHERS

In so far as the "big ranchers" are characterized by greater wealth and a more sophisticated life, they are admired and envied by the Homesteaders. The prestigeful rancher has become a role-model for many of the young boys in Homestead. On the other hand, there has been serious "rancher-nester" [10] conflict over land for the past two decades, which has generated deep feelings of mistrust and even hate by the Homesteaders for the "big ranchers" who are said to "act as if they were lords of the universe." These feelings reach their most intense form in relationships with the "rancher on the south" who in 1929 owned less than ten sections and today controls approximately 150. This rancher appears to harbor a particular hatred for the "nesters" and would like to see all of the Homestead region become ranching country again.

The "nesters" define the Taylor Grazing Act and the Bureau of Land Management of the Department of the Interior as political instruments utilized by "the big ranchers" to keep "all the land for themselves," to stifle agricultural activities and to keep the Homesteader from making an honest living. The "nesters" also point to the fact that the overgrazing on the part of some of the ranchers is damaging the land. The "ranchers," on the other hand, find a coin-

GROUP SUPERIORITY AND INFERIORITY 137

cidence between their interests and "public interests" and point to the damage done when the native sod is plowed up, the native forage is ruined, and the land blows away. They feel strongly that in the long run the land is suitable only for grazing livestock. While the Homesteaders have more votes, the ranchers have more political and economic power, especially as channelled through the Sheep and Cattle Growers' Associations which have emerged as potent political forces in the state as a whole.

It is clear to the less impassioned observer that the truth lies somewhere between the extreme positions taken by the rancher and the Homesteader. In sheer geographical and economic terms (regardless of social considerations which will be discussed later), it is apparent that it has not been to the long-range public interests in soil conservation to plow up all the land that has been placed under cultivation. On the other hand, it is equally evident that some of the land now controlled by the ranchers could be selectively cultivated with no long-run damage to the land. Goodsell points out that a sensible long-range economy for the area could be built around a combination of careful farming and livestock grazing, and he provides ample data to demonstrate that greater incomes would result from such an arrangement.[11] This is no consolation to the prospective Homesteader who arrived too late to stake out a claim, or to the Homesteader who controls insufficient land for livestock raising. The facts mean even less to the old-time rancher who is wedded to a way of life revolving around cattle raising and who views farming activities with contempt.

Recently a few Homestead families have established friendlier relationships with the ranching families, but the essential ambivalence still runs so deep in the community that these few families are subjected to severe criticism for "playing up to the ranchers."

RELATIONSHIPS WITH MORMONS

The Homesteaders' contacts with Mormons on the Plains of Texas were limited to acquaintanceships with a few Mormon missionaries who were assigned to the Texas region. For the most part, the Homesteaders held the familiar stereotypes of Mormons as a people who had a "peculiar religion" and "more than one wife."

Contacts in New Mexico were initiated when the Homesteaders traveled through Rimrock on their way to Gallup. At first there was a sense of relief felt by the Homesteaders when they found that they were not the only "white folks" in the area. Since Rimrock is more distant than Tapala, however, contacts have not been as frequent with the Mormons as with the Spanish-Americans. Baseball games were scheduled between the Rimrock and Homestead community teams; the two communities attended each other's dances and rodeos; and eventually there were three cases of intermarriage. It was obvious that the Homesteaders would like to believe that the Mormons are the same type of rural people with the same values and that they are social equals. But Rimrock has an older, established history in the region, higher material standards of living, and more community activities — qualities which make the Mormon community "superior" in the Homesteader's mind. On the other hand, the Mormon religion has continued to be regarded as most peculiar by the Homesteaders. It was a shock to discover that the Mormons dance in their church and that they open and close their dances with a prayer. In recalling the first time he attended a Mormon dance, one of the Homesteaders stated:

> We didn't know anything about Mormons. Never been around them. Well, we went to that dance and directly they said a prayer. All of our women were sitting around talking, yackety-yackety, and somebody tapped them on the shoulder and said, "They're praying." And you should have seen the look on those women's faces when they saw them Mormons praying to start a dance!

Some years later the Mormons began to make missionary visits and give sermons on Mormonism once a month in the Homestead Community Church. For a time the Homesteaders listened politely, but in recent years most people leave after Sunday School and do not remain to hear the Mormons. Nor have any converts been made among the Homesteaders. As of 1951, my informants were unanimous in expressing the point of view that "All of the people in Homestead think that Mormon religion is crazy." The Mormons also tend to be "clannish" and not as "friendly" as the Homesteaders consider themselves to be. For their part, the Mormons regard their own way of life as distinctly superior to that of the Homesteaders. Some will admit that the Homesteaders have the virtue of being "more friendly" and of "mixing more with others," but Homestead is generally considered a "rough" and in some ways "immoral" community, especially because of the drinking, smoking, and fighting

GROUP SUPERIORITY AND INFERIORITY 139

(particularly at dances) that takes place. They also feel that Homestead is disorganized and that the churches are not doing what they should for the community.

Thus, while relationships among the Homesteaders are markedly equalitarian in nature (the community is split more into various factions than into social classes), the village as a whole has come to occupy a position in a larger prestige hierarchy comprised of the neighboring villages and more distant cities in New Mexico. In this more encompassing hierarchy, the Homesteaders regard themselves as superior to the "Mexicans," Pueblos, and Navahos — all of whom they regard with contempt — but manifest orientations of inferiority and insecurity toward "city folks," and regard "big ranchers" and Mormons with ambivalent feelings of admiration and envy, tinged with mistrust and hate in the case of the ranchers, and with contempt for their religion in the case of the Mormons.

7

The Atomistic Social Order

The Homesteaders' orientation toward social relationships is premised upon a "rugged individualism" which presents a stark contrast to the hierarchical emphasis in Spanish-American culture or the coöperative accent of Mormon culture. In Homestead each person is expected to be independent and self-reliant, and to have an equal voice with every other in community affairs. Coöperation on a community level occurs, but the primary emphasis is upon the individual and his immediate family as the basic unit of social organization. The Great Words, "Freedom" and "Democracy," are given meaning in terms of this individualistic value-orientation. "Freedom" to a Homesteader means being a free agent to buy and sell as he pleases in the economic market, being free of "guv'ment interference" and of the oppressive restrictions of "the law." "Democracy" means independence; "a man likes to have the world by the tail," he likes to "have his own farm" and to "be his own boss."[1]

Interpersonal relations are strongly colored by a kind of competitive "fussin' and feudin'" in which families manifest great envy of one another in the struggle for wealth and prestige within the community. The Homesteaders, living on their widely separated "places," do not ignore one another, as it would be possible for them to do. On the other hand, they do not coöperate in community affairs like a hive of bees. Rather, they interact with one another, but the interaction has a "feudin'" tone to it except when a crisis of sufficient magnitude arises to stimulate a temporary and informal coöperative organization by means of which they deal with the problem. Once the crisis is over, the "normal" feuding and competition are restored as the basic way to proceed in living.

In the previous chapters on the influence of other major value-orientations upon the course of events in Homestead, it has been apparent that this frontier value-stress upon "individualism" permeates much of the culture of the community. The consequences of

No Stained Glass Windows — but Services Take Place Here Every Sunday

Modern New Mexico Is the Land of the Pickup Truck

Homesteader Houses in the Pinyon Trees

THE ATOMISTIC SOCIAL ORDER 141

this prime "individualistic" value-orientation (and of other orientations where appropriate) for the over-all social structure of Homestead will be discussed in this chapter.

ROLE STRUCTURE OF HOMESTEAD

A social system may be viewed as a system of roles which persist over time more or less independently of the individuals who fill them. The role is that sector of the individual's action which forms the point of contact between the behavior of the individual and the social system. These roles are defined by the value-orientations of a culture and may be conceived of as containing expectations and responses to these expectations which organize the reciprocities in the interaction systems of the individual members of the social group.[2]

These roles are found in articulated clusters in larger structures which crystallize around critical foci in the social system and in the human situation. We may conceptualize and describe related roles in an *occupational structure* (concerning the instrumental problems of the use and allocation of human resources in the productive process); an *age-sex structure* (geared to the biological facts of age and sex); a *kinship structure* (emerging from the facts of biological relatedness); a *territorial structure* (emerging from the fact of territorial contiguity in residence); an *educational structure* (emerging from the necessity of transmitting cultural patterns to the younger generations); a *political structure* (concerning the integrative problems of the allocation of power, the control of force, and the development of leadership); and a *religious structure* (which in the broadest sense concerns both cognitive and emotional adjustments to the nonempirical aspects of human existence). Other structures could be added to this framework, but those listed are to be regarded as the relevant ones for the present inquiry.

THE OCCUPATIONAL STRUCTURE

There are two general roles in the occupational structure of Homestead: the "bean farmer" and the "service person." It should be understood that these are general labels, the former referring to farm operators whether the farmer only raises beans or concentrates more on livestock, and the latter to townspeople, including operators of stores, schoolteachers, mechanics, who provide the economic, educational, and other services in Homestead center.

The primary content in the role of the bean farmer may be described in terms of four related role-expectations: (a) He is expected to have his own farm (or "place"); or, if he rents the land he should strive toward buying his own. (b) He is expected to be self-reliant in the operation of his "place"; this includes minimum dependence upon labor outside the nuclear family, and a stress upon "being your own boss" — a value-pattern which is not only found in this connection but pervades much of the social structure, and is always mentioned by the Homesteaders as one of the principal reasons for living in Homestead rather than doing wagework in a city. (c) He is expected to exhibit "free market behavior" in deciding what to plant, how much to plant, and how to dispose of his crops and land in such a way as to maximize his profits. (d) Finally, the role contains a stress upon future personal achievement, measured principally by greater bean production, accumulation of more land, purchasing automobiles and other material goods in competition with other bean farmers.

The fundamental relationship of the nuclear family to the "place" has important effects upon community life. Up to a point the rural "place" is a functional analogue of the urban "job" in providing the Homesteader family with a structurally strategic role in the social system. To own or rent a "place" means that one also has a "place" or status within the system (and from this point of view the structure of the community might be described as a system of "places"); to be without a "place" means that one is without genuine status in the community. This is true whether one is a bean farmer and has "a farming place" or is a businessman in Homestead center and has "a place of business."

The relationship of nuclear family to the farm is also important in that Homestead has a scattered settlement pattern, rather than the pattern of compact villages characteristic of the Mormons, Pueblos, and "Mexicans."[3] The Homesteader wonders why the "Mexicans" or Mormons want "to live all bunched up in those villages." In Homestead the land is laid out in sections, and roads follow the section lines. The Homesteader thinks of his community in terms of these sections which are translated into "places" when they become farm homes.

The scattered settlement pattern means that each nuclear family lives in physical isolation from other families and is the basic unit

THE ATOMISTIC SOCIAL ORDER 143

of social interaction in the community. Only when you "visit your neighbors or your kinfolk" or "go up town" do you come into contact with other people. Each of the scattered "places" has a history which is known by all in the community. Even a ten-year-old boy can tell you that "that place was homesteaded by X, traded to Y in 1940, and is now owned by Z." In addition, both the general characteristics of "places" and the particular characteristics of a given "place" are well known.

Transportation between "places" and into town is now almost entirely by automobile. The Ford, Chevrolet, or GMC "pickup" (a half-ton truck) is the most popular vehicle and is used for business and pleasure. Much "trading" around takes place with the tractors and automobiles, and these vehicles also acquire a history that is familiar to every person. A ten-year-old can also tell you that "that 1942 Chevy there was bought second-hand by X in Albuquerque in 1946 jest after he come back from the war. He traded it to Y for three cows and calves. Then jest last week Y traded it and a quarter-section of land to Z for Z's 1950 Ford."

The car becomes an extension of the personality in sociometric terms. Everyone knows each family's cars (and tractors) by sight, and it is not necessary to meet a person face-to-face to know what visiting and traveling around the community is taking place. A farmer can merely glance up from his work in the fields, see a car passing along the road, and tell you: "there goes X up town for his mail"; or "that must be Y goin' over to visit his brother." The presence of a "stranger" in town is also known as soon as a strange car appears. Needless to say, these facts have important implications for social control in the scattered community of Homestead. Gossip can be initiated on the basis of information that a certain car was seen driving into a certain "place."

In the case of the "service" roles it is felt that these positions exist for the benefit of the bean farmers who do "the real work" of the community. While it is legitimate for them to pursue their individual interests in maximizing profits, they must "be nice to the public" and continually validate their positions. Although these roles are recognized as less desirable, less prestigeful than farm operating and having "your own place," some are more desirable than others. The mechanics, the schoolteachers, and the postmistress are regarded as having indispensable functions, while the storekeepers (and the bartender) are considered less important and are often defined as persons who tend to "overcharge and cheat the public." Finally, these roles are generally characterized by a serious conflict between universalistic and particularistic standards in this small community where everyone is personally known. In the business establishment there is the problem of standard versus reduced prices for "kin folks" and "friends" and the problem of credit and collection of bills from people who have particularistic claims as "kin folks," "friends," or "neighbors." A number of earlier stores "went broke" when the owners could not collect on bills, and a few families still possess "whole suitcases full of unpaid bills." One of the mechanics reported that if he had his choice, he would move to a community where he had no relatives and no friends and would start all over again. In the case of schoolteachers, the same type of problem exists in the particularistic pressures for differential treatment of children from families who are "kin to" schoolteachers.

The different types of "service" roles carry particular role-expectations. The mechanics are defined as persons who, like the farmers, "do *real* work." They manipulate and repair the technological apparatus and "get greasy and dirty" in the process. Their places of business are also the meeting places of the men's loafing groups which they are expected to encourage by providing seating space for the loafers.

Since a post office is not only an institution for the dispatch and receipt of mail, but also a symbol (for the Homesteaders) of an "up-and-coming community," the role of postmistress is also a prestigeful one. The influence inherent in her functionally specific role of handling the United States mail is added to by her more diffuse role as the collector and dispenser of information on intra-community affairs. Indeed, the information a person gets each day at the post office far surpasses in quantity the information he sends or receives via the mail. There is not much loafing as such in or outside the post office (except when the daily mail does not arrive on time), but there is almost always some exchange of information between the postmistress and the person who picks up the mail. In this way, the postmistress controls and passes on more information about the movements of people, the conditions of the roads, and where it did or did not rain in the community, than any other individual in Homestead.

The schoolteachers and "preachers" are strictly "white-collar"

THE ATOMISTIC SOCIAL ORDER

workers, and although there is some feeling that these are "soft jobs," there is prestige attached through the recognition that these positions require superior education (except in the case of "preachers" in certain denominations who simply "got the call"). A special role-expectation is also present in the form of behavior restrictions which apply only to the schoolteachers and "preachers"; they are expected to be moral exemplars and are not permitted to drink, smoke (in the case of women teachers), or play poker.

The storekeeper occupies the most difficult role of all in the occupational structure. For, while managers of this kind of small business enjoy respect and prestige in larger American communities, they are generally mistrusted in this small farming community. Part of this attitude derives from the particular fact that the large general store is currently operated by a brother-in-law of a "big rancher." However, the evidence on attitudes toward the other stores in Homestead, current and defunct, indicates a more generalized mistrust of storekeepers. The roots of this feeling can be found in the nature of the role in this particular situation in which two aspects appear to be critical: (*a*) the Homesteaders' definition of a storekeeper as a man "who don't really do any work"; and (*b*) the conflict between universalistic and particularistic standards in price and credit arrangements. In this situation another important economic factor is that incoming goods and outgoing products of the community are handled by separate business establishments. In other stores in the area (Rimrock, for example) a single storekeeper sells goods to the community and purchases livestock and crops from his customers. In these transactions he makes a two-way profit and enjoys a greater economic margin in facing problems of price and credit. In Homestead, beans are purchased only by the operator of the bean warehouse and incoming goods are sold only by storekeepers, who must attempt to make a living through a 20 per cent markup over wholesale prices on general merchandise. If the storekeeper makes too many particularistic concessions, he "goes broke"; if he makes too few concessions, he becomes the focus of community hostility; and in drought years, he suffers along with the bean farmers.

THE AGE-SEX STRUCTURE

The Homesteaders recognize the following age categories: "babies," "kids" (or "young-uns"), "teen-agers," "married folks," "old folks." "Babies" are ordinarily infants in arms, but the term also applies to toddlers. "Kids" are children up to age of puberty; they are expected to "play and have a good time," although there is early stress on independence, mastery training, and becoming economically useful around the house or farm.[4] At the age of five to seven, boys are expected to begin helping with daily chores (chopping wood, bringing in the milk cows) and with hoeing in the fields, and it is not uncommon to see ten-year-olds driving automobiles and huge four-row tractors in the fields. Small girls of five to seven are expected to begin washing dishes, cooking, and cleaning house. By the age of ten or eleven they can cook complete meals, including cakes and pies. Both boys and girls are taught to be respectful and to exhibit deference to adults, as evidenced by the ubiquitous use of "Sir" and "Ma'am" in addressing older persons. Boys and girls are believed to be markedly different in basic nature, however. Boys are expected to be "mean and mischievous" and to need a great deal more control than girls, who should be "good most of the time."

After puberty there is the teen-age or high-school period in which some "wild behavior" is considered inevitable. There is also a curious reversal of sex role-expectations at this time in that the teen-age girls are defined as much more energetic and more mischievous than the boys. It is true that boys are likely to "sow wild oats" while "nice girls" are not permitted this liberty. But while the boys engage in their activities relatively quietly, the girls form a well-organized group which is loud and boisterous and a source of continual discipline problems in the high school. It is during this age-period that the "high-school crowd" engages in such activities as "chicken roasts," with chickens "stolen" from some farm, and in the familiar pattern of dating, dancing, petting, etc. In so far as we could determine, however, petting seldom leads to complete sexual intercourse, and there have been few cases of premarital pregnancies and "shotgun weddings."

The full-blown "youth culture" with patterns of "hot-rodding," etc., is absent in Homestead. Instead, the adolescent period is usually brought to an abrupt end by an early marriage, often before the two young people have finished high school. Parents make an effort to delay the marriages until after high school (there is almost never any pressure or expectation by the parents that the young will finish college before they get married), but the attempts are feeble and the marriages usually take place with the approval of the com-

THE ATOMISTIC SOCIAL ORDER 147

munity. Marriage choices are believed to be matters of an individual boy and girl falling in love, and the role of the family in such decisions is minimal.

Following marriage, young couples are defined as "married folks" and take their place with the others as pillars of the adult social structure. The distinction sometimes made between "young married couples" and "old married folks" is functionally unimportant except in such matters as active participation in dances and other strenuous activities which gradually declines with advancing age.

There are no "old maids" in Homestead, but there are five widows and five "bachelors." A "bachelor" is defined as any man over marriageable age who "batches it," i.e., lives alone and does his own cooking and housekeeping. Actually all the "bachelors" have been married, but have lost their wives through death, separation, or divorce.

After the age of about sixty, people become "old folks," but are expected to continue to work and support themselves, if possible, as long as they live. Since Homestead was settled by migration of predominantly young adults, there are very few "old folks" in the community.

As in all human societies, the social division between the two sexes is fundamental. In general, there is a clear-cut economic division of labor within the nuclear family. Wives are housekeepers and mothers (with less of the companion or glamour-girl role-expectation found in urban areas); husbands are breadwinners and fathers. The women do the cooking, house cleaning, laundry, sewing, canning, and caring for the children; the men do the farming, the house building, the chores, water hauling, the wood hauling and chopping, and repair the machinery. In many situations, however, they may perform tasks and assume responsibilities of the other sex. During the busy harvest season women often work in the fields; water hauling sometimes has to be done by the women; wood chopping is ideally men's work, but when the husband has failed to chop enough, the wife sometimes has to chop and carry in wood to finish preparing a meal. Men sometimes help with the cooking, and although child care is primarily a woman's responsibility, the husbands often help, especially with the boys. In general, the balance between the two sexes is equal in power terms,[5] but is unstable and productive of much cross-sex strain in the social structure.

This cross-sex strain arises from a combination of a structural feature of the social system and a psychological feature of the socialization process. The structural feature consists of the fact that although the husband and wife are expected to work closely together in the operation of the farm, a large amount of permissiveness of independent action is granted both sexes in community affairs. In contrast to Spanish-American women, for example, Homestead women may visit around the community by themselves and may drive into the city on shopping trips without their husbands.

The psychological feature is essentially an aspect of what Phillip Wylie terms "momism" and concerns the psychological dependence of the small boy upon the mother and the husband upon the wife. Thus, while men are self-reliant and independent in the area of economic responsibility, manipulation of technology, etc., they are emotionally dependent upon "mamma." "Mamma" is first the mother, later the wife, and is defined as the "moral person" who curbs and controls the unruly impulses of the "bad boy." The evidence for "momism" lies in the fact that small boys are defined as "mean and mischievous" and in need of much control by the mother. Later one observes adult men being mature, self-reliant, and independent in the operation of their farms, but calling their wives "mamma" and behaving "like grown-up little boys," especially in situations where they are aggressively striving to assert their masculinity, but not quite carrying it off.

For example, in poker games, in drinking parties at the local bar, and in drinking groups outside dance halls, men almost always refer to their wives as "mamma." When a man is playing poker, he will say, "I'll bet mamma won't let me go to the dance tonight if she finds out I've been playin' poker." "Mamma" may come to "drag" the man home from the bar if she thinks he has been drinking too much and too long. In a drinking group outside a dance one night, one of the men saw his wife coming outside and shouted, "Here comes mamma!" — whereupon the men grabbed their bottles of liquor and ran and hid behind the schoolhouse to continue their drinking.

Growing out of these structural and psychological features of the sex roles in Homestead with the conflicts between a high evaluation of a certain kind of "rough and tough" masculinity on the part of the men and a high evaluation of women (an "equality of sexes"

THE ATOMISTIC SOCIAL ORDER 149

ideology), is a form of symmetrical schizmogenesis between the two sexes found both in husband-wife relationships and cross-sex relationships in community social events. In the husband-wife relationship this takes the form of "doing and saying mean things" to the other over a period of several days. For example, the husband may be spending too much time playing poker, and the wife will retaliate by "cussin' him out" and going to town to spend his money. The husband may then spend a day at the bar drinking beer. Finally, it may reach the point where the two are not on speaking terms. In fact, there are cases in our field notes of husbands and wives who did not speak to each other for four days at a time.

Not all families, by any means, go through these periods of argument and strife, but it has been observed in at least attenuated form in fifteen families in Homestead. It has also become a culturally patterned feature, as evidenced by the fact that one observes husbands and wives bickering publicly and going through all the forms of this schizmogenic process when our family case studies indicate that relations are harmonious at home and in private.

This "battle of the sexes" is not confined to specific husband-wife relationships but also occurs in larger community affairs. This may be illustrated by the sequence of events in the late spring of 1950. Most of the young wives (age twenty to forty) had been participating for several months in bimonthly meetings of the "Bridge Club" while their husbands took care of the children these Saturday afternoons. In April the husbands organized a "Roping Club" (to practice for the summer rodeo) and deliberately scheduled their first meeting on the same afternoon the Bridge Club was to meet. The move generated much hard feeling, and the wives retaliated by organizing a "Women's Baseball Club," scheduling the first game on the next date the Roping Club was due to meet. So it went until this sequence of events was interrupted by the arrival of the busy farming season. The same kind of process occurs during dances when the men go out to drink "salty dog." The women's patterned reaction is to become "mad," threaten to leave and never come to another dance. The men in turn react to this threatened female dominance with more drinking in order to assert their masculinity and to protect their status in the eyes of their fellows and themselves. Again, a schizmogenesis occurs in which men on one side and women on the other tend to develop into solidary factions around the ostensible issue of drinking.[6] The evidence clearly points to these cross-sex conflicts as one of the significant structural strains in the Homestead social system.

THE KINSHIP STRUCTURE

The fundamental kinship unit in Homestead is the isolated nuclear family in which the crucial bond is the one between husband and wife. In 1950, fifty-one of the sixty-one households in the community consisted of nuclear families with no grandparents, aunts, uncles, or other relatives attached. The other ten households consisted of (*a*) four "bachelors" and one widow who lived alone in separate houses; (*b*) one divorced man who lived with a married daughter and her family; (*c*) four households composed of widowed mothers and their unmarried adult sons.

Despite the cross-sex conflicts which were described above, the nuclear families remain the effective economic, social, and emotional hubs for the individual Homesteader. I have already sketched the relationship of the nuclear family to the "place," the patterned division of labor between husband and wife, and the role of the children in the family. It should be added that there is little record of philandering and extramarital affairs in Homestead even with the opportunities which arise from the community's scattered settlement pattern. In the normal course of events a husband may find himself alone with another's wife when he pays a visit to a neighboring family and finds that the husband has gone to town. Unlike Spanish-Americans, who assume that sexual relations are likely to take place unless chaperons are present, the Homesteaders assume the existence of internalized controls in adult people which will ordinarily be effective in preventing extramarital affairs. In addition, our evidence indicates that husband-wife sexual relationships are (with a few outstanding exceptions) reasonably harmonious in Homestead, and there appears to be little temptation to seek (or to develop fantasies about) illicit intercourse with other partners.

When adultery does occur and is discovered, the typical pattern is for the husband to "beat up" the man who slept with his wife and to institute divorce proceedings. These facts underline the strategic importance of the husband-wife relationship in the social structure and the reluctance of a husband to share his "mamma" with another man.

THE ATOMISTIC SOCIAL ORDER 151

Beyond the nuclear family, which is referred to simply as one's "family," are the "kin folks," who include adult "brothers" and "sisters," "grandparents" or "grandchildren," "aunts" and "uncles," "nieces and nephews," and "cousins."[7] These terms are applied to biological relationships, but are extended to purely sociological relationships in the case of "aunt" and "uncle," which are commonly used by younger people in addressing older, unrelated persons in the community. Beyond the fact that married brothers and sisters are expected to visit one another and to coöperate to some extent in economic transactions, and that the grandparent-grandchild relationship is expected to be a warm and permissive one, there is little discernible patterned structure to these sets of extended kinship connections. Certainly there is nothing comparable to the highly patterned kinship behavior existing between particular sets of extended relatives in nonliterate societies.

Although the nuclear family is the solidary unit, and the one relied upon in times of stress and tension in inter-personal relations, there are five extended kin groupings in Homestead which are more highly organized and somewhat more intra-coöperative than the others. In each case the structure of the unit is the same: an older father and mother plus one or more of their adult married children and the grandchildren. The members of these kin groups live in separate households and manage separate farms as nuclear families, but in a number of ways they coöperate as a single unit. In actual numbers, they comprise fifteen of the sixty-one households and fifty-seven of the 232 persons in Homestead, or about one-fourth of the community.

The most conspicuous of the large kin groups, and the only one which is called a "clan" by the Homesteaders, is composed of an older father and mother, four of their married children, and the grandchildren, plus one additional family which has a more peripheral connection by virtue of being "cousins." It is the only kin group in Homestead with a published family history, which goes back to Revolutionary days in Virginia and is traced through a westward family movement to Tennessee and on to Oklahoma. It is the only family which holds formal "family reunions" and large family picnics.

There are a total of six households and twenty-seven persons involved directly in this kinship unit. As the nucleus for a large and fairly stable faction throughout the history of Homestead, there is no doubt that this kin group tends to dominate the formal social organization of the community. They control the Community Church, the Farm Bureau, the post office, and the school system (three of the adult children are schoolteachers). However, their economic position is not particularly strong: they control less than nine sections of land and only two of the six families have incomes of over $5000; the others have only average incomes. Moreover, without the supplementary incomes from schoolteaching and the post office, their incomes would be substantially lower.

A second kin group is composed of an older mother and father, two of their married children, and the grandchildren, with a total of three households and ten persons. This group operates two farms and the combination store and *café*, and also forms the nucleus for a second but less stable faction which dominates the Baptist Church. Their economic position is even less secure since they control only one and one-quarter sections of land and have no salaried positions in the community.

The three other extended kin groups are composed of an older father and mother (in one case the father died during World War II) and one married child and grandchildren; in each group there are two households and five to ten persons. One group controls almost three sections of land; the other two have between one and two sections each.

The areas of expected coöperation in these kin groups are: (*a*) economic — exchange labor in harvesting, butchering, and housebuilding, and mutual assistance in the buying and selling of land and other property; (*b*) the care of children — usually at another's house while their parents are at a dance, or making a trip to the city; (*c*) inter-visiting and inter-dining, particularly on Sunday after church.

In these coöperative activities the large "clan" is the most successful and therefore is singled out for special comment by the rest of the Homesteaders. It is evident, however, that even in the "clan" the reciprocal relationships are far from harmonious. The execution of larger coöperative endeavors is constantly accompanied by intra-kin competition and conflict which frequently result in a retreat back into the nuclear family for support. In the other kin groups the coöperative endeavors proceed on even more tenuous bases. Nevertheless, these extended kin groups are small islands of coöperation

THE ATOMISTIC SOCIAL ORDER 153

within the larger structure characterized by an individualistic commitment to one's own family and "place."

The forty-seven nuclear families who are not associated with the five extended kin groups described above are not all without "kin folks" in Homestead. More than half of these families are "kin to" one or more other nuclear families in the community. There is some inter-visiting and inter-dining along kin lines, but no economic or child-care coöperation. In the case of three families in which there are two or three married siblings present in Homestead, arrangements are more on an individual nuclear family basis typical of the larger community.

Within the circle of "kin folks" there is the special category of "in-laws." They are sometimes called "the out-laws," especially by persons married into one of the larger extended kin groups, which indicates the strained relationship when one marries into a larger family. On the other hand, the "in-law" relationship is often used in a positive way to establish rather tenuous kinship connections in the community. One marriage between two families is conceived as relating all the members of the two families — not merely the two persons directly involved in the marriage — and thus enlarging the circle of "kin folks." Homesteaders like to think of themselves as being "kin to just about everybody" via these affinal connections. These connections are fundamentally different from the kinds of kinship relationships existing, for example, in the Mormon village of Rimrock where there are a multitude of kinship connections of aunt, uncle, and cousin. It is possible that the Homesteaders' tendencies to relate large families merely by affinal connections and to extend "aunt-uncle" terminology to nonrelatives, are attempts to bring everyone within the known circle of kinship, although the biological connections in this small personalized community are tenuous as compared to Mormon, Spanish-American, or nonliterate communities of comparable size.

THE TERRITORIAL STRUCTURE

The most general role in the territorial structure is that of being a resident of Homestead, of living within the geographical bounds of the community. Within the area of the community there are two "neighborhoods" of some importance which have historical roots in the fact that they once hoped to become independent communities.

There was once a school in one area, and a school and post office in the other. Later, these were shifted to Homestead center, but the people in these two neighborhoods still tend to feel that they are living in special areas. There is still some economic reciprocity which is organized on a larger neighborhood basis, such as exchange work for harvesting.

A more important role based upon territorial contiguity is that of being a "neighbor." "Neighbors" are expected to visit each other, be able to borrow food, tools and other items back and forth, and engage in some exchange work. The "change work" takes the form of "you hep me and I'll hep you," and is today found to be most important in the tasks that require the most labor: setting up a windmill, butchering, house building, and harvesting. There is much less "change work" now than there was in the 1930's when the beans were harvested with the threshing machine and coöperative groups worked several neighboring farms. The present use of the combine gives the farmer more independence in his harvesting operations. Furthermore, geographical proximity in itself is no longer as important as in the days before rapid transportation by automobile. Finally, the factors of kinship, religious affiliation, and factional alignment often tend to override these local neighborhood connections among the Homesteaders.

The pattern of reciprocity, or "paying back" work or other favors, is immediately noticeable and very important in Homestead. There is an almost compulsive tendency to repay a favor as soon as possible; the Homesteaders do not like to be dependent upon or "in debt to anybody" for favors done — another important manifestation of the stress upon self-reliance and individualism.

THE EDUCATIONAL STRUCTURE

The task of socializing the young in Homestead is divided between the family (as described above) and the school system in the village. The school is a concrete expression of the value placed on "getting an education" so that one can become a "success" in the future. This value has, of course, a formal codification in the state law that children under sixteen must attend school, but the Homesteader has always been a strong supporter of education, which now means a high-school education. A half-dozen people in the community have been to college, and they and a number of others plan to send their

THE ATOMISTIC SOCIAL ORDER 155

children to college. For most of the Homesteaders, however, "high school is enough." Earlier generations received comparatively little education, and there are many people in Homestead who only had a grade-school education or less. They are sensitive about their lack of education; many prefer not to mention it and give evasive answers when asked how many years they went to school, but as parents, they are determined that their children will have more education.

The school is more than a place where the children acquire an education; it is the most important center for communal activity in Homestead, not only because almost everyone has children in school and feels he should attend and support school functions (programs, graduation exercises, meetings to promote the hot-lunch plan, etc.), but also because the schoolhouse is the largest community building. It is here that Saturday night dances, political rallies, weekly movies, and meetings of all kinds are held. Churches, stores, and other buildings stand for special interests, but the schoolhouse is a *community* building and a *community* symbol.

THE POLITICAL STRUCTURE

Since the village of Homestead is unincorporated, there are no town officials and no state or county officials except for a man who is part-time deputy sheriff and game warden. His principal occupation is the operation of the shop that specializes in the repair of automobiles. It was some time before the writer realized that the role of deputy sheriff and game warden carried the principal expectation that *nothing* was to be done by this "official" representative of "the law" either when state laws or regulations were violated or when there was a situation of violence or disorder in the village. Indeed, action by outside authorities in the affairs of this small community is generally regarded by the Homesteaders as unnecessary and unwelcome interference. This is particularly true in the violation of the game laws, which are felt to be instruments for protecting the deer for "city folks" instead of allowing the Homesteaders to use them for meat. As one old-timer expressed it, "God put the deer here for a purpose . . . to be used for food by the poor folks."

In the early 1930's the Homesteaders had no sheriff or Justice of the Peace in the community. They took their legal conflicts (which mainly consisted of "feuds" between nuclear families or "fights" between a husband and wife) to the Justice of the Peace in the neighboring Spanish-American village or to the Justice of the Peace in another homesteader community (now defunct) twenty miles to the east. Florence Kluckhohn has described how the Spanish-American Justice of the Peace was so distressed by having to deal with all these Homesteader family squabbles, that he eventually resigned from his position in 1936.[8]

In the late 1930's Homestead had both a deputy sheriff and Justice of the Peace. Not only were they completely ineffective in matters of law and order, but they also actively participated in various "feuds." It was during this period that the Homestead Justice of the Peace attacked the "preacher" with a knife one day at church. The case is best described in the words of one of the old settlers:

> One time the Justice of Peace got drunk and chased the preacher around the Community Church with a knife. I and two others got together and went down to see him after this happened and told him he'd have to pay a fine. The deputy refused to arrest him and take him to court, so I went down to get him myself and took him over to Tapala [the Spanish-American village] but they didn't have any Justice of Peace over there then. So then we took him down to the Justice of Peace at Ventana. A number of cars of people from Homestead went along. A bunch of the women piled up enough charges agin' our Justice to hang him. When they rattled off all these charges, the Justice at Ventana said this would be too much for him to handle, and they'd have to take him to the county seat. But no one wanted to take him all the way to Los Lunas, so the women withdrew most of their charges and jest charged him with disturbin' the peace. So the Justice fined him $5.00 and everybody went home. Then after they left, the Justice filled up my pickup with gas and gave our Justice the $5.00 back!

In the early 1940's the men who had filled the positions of deputy sheriff and Justice of the Peace moved away from the community, and the positions were not filled again until 1946 when the present deputy sheriff was commissioned by the county sheriff. The county officials tried continually to establish a deputy in Homestead, but no one would accept the position. One old settler reported:

> I was elected for deputy twice, but I never filled out them papers. I always said to those fellows down at Los Lunas, "If somethin' happens, we'll bring them in ourselves. I don't want to make no enemies by bein' deputy sheriff." Most of us around here felt like I did, that if anybody needed arrestin', one or two of us would get together and take them in . . . That man was appointed because nobody else would take it.

THE ATOMISTIC SOCIAL ORDER 157

Since the present sheriff was commissioned he has made no arrests, nor is he expected to. A number of years ago, when he was a "newcomer" to the community, he appeared at a Saturday night dance with his badge and gun and attempted to stop a fight. The people stopped fighting and turned on the sheriff "and told him to go home and mind his own goddam business." He did leave and has never again appeared at a dance. There have been other occasions since 1946 when "the law" might have taken a hand in the situation. For example, in 1947 one of the Homesteaders was found dying in his home with a bullet wound in his head. The state police or other authorities were not notified; neither were the services of the local deputy utilized. Again, a group of the old settlers informally took control of the situation. They summoned a Justice of the Peace from a neighboring community and formed a coroners' jury for the inquest in Homestead. The case was pronounced "suicide," and the Homesteader was buried in the local cemetery. Today community opinion is still divided as to whether the case was suicide or murder, although it probably was suicide. At any rate, nothing further has been done about it.

Further evidence of the resistance against calling in the official county and state authorities ("the outside law") to deal with violations, is in the ubiquitous practice of hunting deer out of season, which is done and considered proper by even the "moral pillars" of Homestead, such as the principal of the school, as well as the game warden himself.

Indeed, there are only two points at which there is rigid compliance with the county and state laws: (*a*) marriage and divorce, and (*b*) the ownership and transfer of land and other property, such as cattle and automobiles. These facts underlie the crucial significance of the nuclear family and of the institution of property, especially the ownership of land, in the social structure. The position of the nuclear family is further buttressed by the strong taboo against adultery. As mentioned above, when cases of adultery do occur (we know of only four cases in the history of Homestead), they are promptly and unequivocally dealt with — first with "fist fights" and later (in all four cases) with divorce.

There is also acceptance of most of the major legal requirements involving the school system and the post office, but in the other areas of community life it is clear that "the law" is resented and regarded by the Homesteaders as unnecessary interference. The feeling as expressed by another of the old-timers is that:

> We've never had anybody do anything around here very bad. In the twenty years in the history of Homestead there's never been anybody who's had to be taken to the county seat for a trial. Of course, we've had a few little fights and a few little fines, but that's all that's happened and I think that's a pretty good record. When these fights happen, we usually just talk to the person and we don't have to go to the state or the county.

The rejection of the law represented by outside authorities does not mean that the Homesteader is "lawless"; indeed, he feels that he is a moral and law-abiding person. Rather, it means that the Homesteader's definition of what constitutes "crime" varies at many points with the county and state legal definitions. Further, it indicates that the controls for deviant and "lawless" behavior are present *within* the community.

The informal controls for disruptive behavior take a number of forms. There is first the perennial gossip directed against this deviant behavior and, correlatively, a marked sensitivity to "being gossiped about" in the women's informal gossip groups and the men's loafing groups. There is a patterned difference in the kind of "gossip" of the two sexes. Women relate incidents that have happened recently in the community and express their approval or disapproval of the behavior. Men more often tell long stories of things both recent and long past, often exaggerating the truth — but this too is part of the pattern. Thus each day the important values which regulate and reinforce the social system are given expression.

If gossip alone is not effective in correcting the behavior in question, or if the offense is more serious and involves two males, a "fist fight" may ensue, especially at dances where there is a great deal of drinking and tempers may be short. This fighting itself does not represent an uncontrolled application of force, but always proceeds in terms of clear-cut conceptions as to what constitutes "fair fighting." In the first place, fights must take place outside the dance hall, away from the view of the "lady folks." Only one man may fight one other man; for two men "to jump on one man" is a strong violation of the fighting code. The writer has observed occasions where "the crowd" (i.e., the male spectators) have actually "beaten up" fighters who did not conform to these basic rules.

Finally, if the violation committed is really a serious one, the situa-

THE ATOMISTIC SOCIAL ORDER   159

tion is handled by a group of the informal leaders in the community who talk to the person involved and, if necessary, take him to a Justice of the Peace in a nearby community for a trial and fine. Summoning the state police or the county sheriff is simply never done.

To provide a concrete example of the operation of these controls, it may be noted that hunting deer out of the legal season (as defined by state law) is not defined as a "crime," but as a legitimate, economic pursuit. The real "crime" is to report someone who hunts out of season to outside authorities. When this happens (as it does occasionally), "all hell breaks loose" within the community and the person who "turns somebody in" is punished by techniques ranging from gossip, through social ostracism, to "beating them up," in the more extreme cases. Furthermore, deer cannot be killed indiscriminately, and to do so means one is a "game hog" whose actions will also be controlled by informal but effective methods.

Informal mechanisms are also present for the handling of crises. For example, when one of the Homesteaders "went crazy" recently and started to kick out the windows of his cabin, the deputy sheriff again appeared thinking this might be a situation which would require his services. Again the sheriff was told to go home and "mind his own business," and the case was taken care of by a group of informal leaders who took the insane man to the state mental hospital.

In positive community action, as well as in the control of disruptive behavior, these informal leaders are the crucial persons in the social structure. There are a number of formal leadership positions in the community; e.g., the officers of the Farm Bureau, the water coöperative, and of the rodeo committee, but these positions are strenuously avoided by the "real" leaders and are now typically filled by "newcomers" in Homestead, "people who haven't learned better yet." The "real" leaders prefer to operate informally by simply talking to people during visits at home or in the loafing groups. Two types of informal leaders are present in Homestead, and may be characterized as "the organizer" and "the exemplar." The first type is constantly busy initiating "behind-the-scenes" action, manipulating people and ideas by a "lot of talk" throughout the community. He is often considered a "meddler" and has less influence than the second type, who never tries to initiate anything, but is a leader "by example." He is the "good farmer" who usually stays home and "minds his own business." Although he may spend time "up town" visiting leisurely with his fellow Homesteaders, his strength as a leader lies in the fact that other Homesteaders come to see him for advice on farming practices, care of machinery, etc.

Finally, there is the role of the "politician" forming the crucial link between Homestead and the county and state governmental offices. Strictly speaking, this "political" role has nothing to do with intra-community governmental and political problems. Rather, it concerns (a) mobilizing public opinion in Homestead for the election of county, state, and federal officials; and (b) particularistic contacts with these officials once they are elected, to insure that Homestead makes "progress" and gets the improved schools and roads the community feels it should have.

The majority of Homesteaders are, of course, Democrats, since their roots are in the South. In 1950 there were ninety-two individuals registered as Democrats, twenty-one as Republicans, and eleven as Independents. Two facts make "politics" important in Homestead: (a) the community is located 177 miles from the county seat where officials have a tendency to "forget" about Homestead's existence; (b) the county government has been controlled by Spanish-American Republicans throughout almost the entire period of Homestead's history. Because of these circumstances, "you really have to scream to be heard" at the county seat and at the state capital. Homestead has had an endless series of committees in the past two decades which were appointed for the express purpose of making trips to Los Lunas and Santa Fe to bring pressure upon the higher officials. There has been a "road committee," a "school committee," a "school-building committee," a "telephone committee," a "well committee," and so on. One of the old-timers commented, "Homestead has more committees than Chicago." Each political party has a precinct chairman and precinct committeemen in the community, and the election officials as well as key members of the various "committees" are drawn from this group of "politicians."

The campaigning in Homestead is done informally by the local "politicians" who speak of their favorite candidates to the loafing groups in the bar, shops, and general store. Then, at some point before the election, the county candidates journey out from the county seat and a formal meeting is held in the schoolhouse during which the candidates speak in their own behalf and on behalf of

THE ATOMISTIC SOCIAL ORDER 161

the state and federal candidates in their political party. This is followed by a "free dance," paid for by party funds, and "free drinks," in the dark outside the dance hall, consisting usually of whiskey dispensed by the candidates or their campaign managers.

After election, the Homestead "politicians" are expected to make trips to the county seat, and if necessary to the state capital, to speak to their "political contacts" about the needs of the community. In describing the leading "politician" of Homestead, one of the old-timers commented with pride that "Jack could go to Santa Fe tomorrow and get in to see anybody, the highway commissioner or even the Governor, and they would listen to what he had to say."

THE RELIGIOUS STRUCTURE

The direct influence of formal religious values and doctrine, except on a few very devout people, is neither wide nor deep in the community of Homestead as compared with other cultures in the Rimrock area.[9] Active religious participation is also markedly less than in the Homesteaders' ancestral home in western Texas.[10] There are two organized churches in the village, the Community (now Presbyterian) and the Baptist, but together they have an average Sunday attendance of only forty-five persons. With the exception of two men who are professed "atheists," all of the Homesteaders are nominally "Christians," and they or at least their parents are affiliated with one of the many Christian denominations. In addition to the Baptists and Presbyterians, there are Methodists, Nazarenes, Campbellites, Holiness Church members, Seventh-Day Adventists, Mormons, Catholics, and "Present-Day Disciples." If asked, most Homesteaders can recite some essentials of the Christian doctrine and theology and add that they are "believers." But aside from the active churchgoers, the Homesteaders are not at all articulate about their religious beliefs, nor do they spend much time thinking about them privately.

The principal discussions of the differences between the various denominations usually center around such points as the mode of baptism. For example, the Presbyterians "sprinkle," while the Baptists "immerse . . . they pitch them plumb in under the water." There is also discussion of the fact that both the Presbyterians and Baptists believe that "once you're converted and saved, you cain't back-slide," while the Methodists believe "you kin back-slide." The

Seventh-Day Adventists have even more divergent doctrines involving a prophecy of the world's doom, a taboo on eating pork (which stands out as very peculiar in this pork-eating culture), and a firm belief that the Sabbath comes on Saturday "because the Lord worked six days and then rested on the seventh . . . and since Sunday is the first day of the week, then Saturday, not Sunday, is the Sabbath." The fact that the Seventh-Day Adventist families stop their farming, marketing, and cooking operations from sundown Friday until sundown Saturday is a matter for special comment by the rest of the community.

Although the Seventh-Day Adventists have "services" in their homes on Saturday, the only other organized services take place in the Baptist and Community churches. When the Community church was organized in 1937, the church leaders said that "anybody can preach in it except Catholics and Mormons." Later it was decided that this was undemocratic and it was ruled that "anybody can preach in it." Thus, for a number of years there was a "Community Sunday School" each Sunday morning, and "preachin'" by a Presbyterian, Baptist, or Methodist preacher, by Mormon elders, and occasionally by others, such as the Seventh-Day Adventists, according to a monthly schedule. But "the Catholics never came," and the Mormons do not come regularly.

The Baptists in Homestead have always objected to the Community church. One of the earlier Baptist preachers commented (when he was asked to participate in the Community church) that he would "rather support a whore house than have a community church in Homestead." The Baptists eventually built their own church (in 1940) and now hold separate services each Sunday. The other denominations also stopped preaching in the Community church, and finally in 1951 the "fiction" was dropped and the Community church became organized as a Presbyterian church.

The Community church is a rectangular adobe building with a pitched roof. There is no cross or sign outside and no pulpit or furnishings inside, except for the hand-hewn pews, to indicate that the building is a church. The "preacher" does not wear a robe or any vestments. There are no collection plates; a hat or folded Sunday-School book is used instead. The only religious articles in evidence are the Bible, the hymn books, the Sunday-School literature, and a map of the Holy Land on the wall. The Baptist church is constructed

THE ATOMISTIC SOCIAL ORDER 163

on the same pattern as the Community church, and other than a recently placed sign on the outside, reading "First Baptist Church," there are no religious appointments. The "preacher" is considered a specialist in interpreting the Bible to his congregation in sermons which are commonly called "preachin'." People say "there'll be preachin' tonight," indicating that a service is to take place.

There are currently three "preachers" in Homestead. One is the father of the large kin group called "the clan," and although he is now "retired," he still teaches Sunday School in the Community church, which has always been dominated by this Presbyterian family. The connection between this family and the Community church is such that on many Sunday mornings there are no other villagers in church — only the grand old man of the "clan" preaching to his children and grandchildren. The Community church now also has a young Presbyterian missionary as a "preacher." He has a "circuit" consisting of four villages and three sawmills, and he appears in Homestead to perform services only on the first and third Sunday evenings of each month. The third "preacher," a Baptist, is a resident in the community and holds services not only on Sunday morning, but also on Sunday evening and often during the week. Both churches hold "Bible Schools" for the children during the summer vacation. In addition to the "preacher" in the Community church, there are the positions of elders, deacons, Sunday-School Superintendent, Sunday-School teachers, pianists, song-leader, and secretary-treasurer.

The nature of the services in the two churches is quite different. While the Presbyterian services are serious and sedate, the Baptist services, especially during "revival" meetings, are accompanied by crying and shouting on the part of the "preacher," and by crying and emotional conversions followed with testimonials by the congregation. The Baptists single out the sins of drinking and dancing for special condemnation, as illustrated by the following excerpts from a sermon on "The Evils of the Dance":

Real ministers of the gospel should do away with dances. I hate the modern dance. And that goes for the square as well as the round or modern dancing. Some people try to give the excuse that dancing is good exercise. That's no excuse at all. They know and you know what dancing really is. And I shall denounce it here and now for what it really is. Dancing is the child of prostitution, the sister to drunkenness and lewdness, and the mother of adultery and murder . . . Nakedness and drunkenness began with the dance and arouse the passions of sex. There was a poll, a secret poll, taken in a college once. Thirty boys were asked to answer frankly the question, "Does dancing arouse evil passions in you?" and each one of these thirty boys answered "yes." Doesn't that show you how evil the dance really is?

It is significant that the non-Baptist Homesteaders responded to this sermon by giving a dance in the schoolhouse on the same night after the revival meeting.

The other denominations have very few members in Homestead. With the exception of one woman of French Canadian ancestry who is a Catholic, the only Catholics in the village are the two Spanish-American women who have married Homesteaders [11] and the local Spanish-American bartender and his wife. The two Mormon members of the community are also women who have married into the Homestead area.

The Seventh-Day Adventists were converted to this sect in 1934 by a wandering "preacher" who stayed in the community for about a year. He moved on after it was rumored that "he was wanted by the FBI" and that he had "run off with another man's wife" and was living with her in Homestead. In 1941, two men who called themselves "Present-Day Disciples" arrived in Homestead and started holding religious meetings. They modeled their activities on the disciples of Christ who worked in pairs and wandered through the Near East holding religious meetings. They did not believe in churches, but held the meetings in homes and accepted meals and clothes, but no money, from the Homesteaders. After three families were converted, the movement came to an abrupt end when gossip circulated to the effect that the "preachers" were having affairs with the wives while the husbands were busy in the fields. The "disciples" moved on to another community, and then "wrote letters back to the people like the epistles of Paul."

INDIVIDUALISM AND FACTIONALISM

There is seemingly a contradiction in terms in stating that on the one hand the Homesteaders are fundamentally committed to an individualistic value-orientation and that on the other hand a system of factions constitutes a basic dynamic feature of the social system. Actually, the two phenomena are not paradoxical, but are related in that the individualistic orientation provides and legitimizes the

THE ATOMISTIC SOCIAL ORDER 165

basic expectation that the individual Homesteader (and his nuclear family) will function as a relatively independent unit in the social system. The larger social units which arise from some common basis are continually under pressure from the stress upon the more individualistic modes of action. A slight increase of inter-personal tension within a larger social unit can result in divisiveness which may then reduce the group in question to the atomistic state which more generally characterizes the total community. These basic features of the social system therefore underlie the shifting factionalism which permeates the community.

For our purposes, a faction may be defined as a group of nuclear families who participate together in social events, including inter-visiting, inter-dining, and "parties," and who function together in intra-community "political" conflicts over control of churches, the school, and associations such as the Farm Bureau, the Water Coöperative, the Rodeo Committee, and the Baseball Club.[12] Economic coöperation may also be present within the factional unit but is not usually expected. Participation in "social events" and coöperation in intra-community "politics" are the key areas of cohesion.

Close analysis reveals that the *particular* factors which account for the composition of any single faction are immensely complicated. The critical factors may be any combination of the following: kinship ties, religious affiliation, neighborhood attachments, political beliefs, equivalency of social rank, personality types (which, for example, underlie mutually supporting role networks of persons with masochistic-sadistic trends in their personality systems), or chance events in the lives of individuals. To explain the structure of each faction in the present scene, one would need to know the life history of each Homesteader. In more general terms, however, we can interpret this complicated factionalism with its shifting alignments as an important aspect of the individualistic orientation.[13] It is evident that factions have always been present in Homestead and that the people expect them to be. No one is surprised when there is a factional realignment, or when some small faction ceases to exist and its members again focus their activities within their nuclear families. It is further assumed that factions will always be "feudin' with one another."

Although we must conclude that factionalism of the type found in Homestead, especially with the continual tendency for realignment and/or disintegration, is related to the value stress on individualism and that a full explanation of the composition and development of a given faction requires knowledge of a vast number of particular factors and events, we can nevertheless trace the relationship of this factionalism to kin groups, to religious affiliation, and to social rank with some clarity.

The most conspicuous relationship of factionalism to kin groups may be seen in the fact that extended families provide the nuclei for all of the large and powerful factions in Homestead. However, factions do not coincide directly with these extended families. The faction always includes nonrelatives, and different nonrelatives from year to year. Moreover, there are many extended families who do not form nuclei for factions; indeed, the members of the families may be distributed among different factions. For example, the largest faction in Homestead has centered around the extended family called the "clan," but has always included additional members who were not related by kinship, and these members change from year to year.

The relationship of religious affiliation to factionalism is also significant in Homestead. In the most general terms, the enormous amount of religious differentiation in Homestead can be interpreted as a function of the individualistic and factionalizing tendencies in the social system. In a culture with a value stress upon independent individual action combined with a "freedom-of-religion" ideology, joining a different denomination becomes one important means of expressing individualism and of focusing factional disputes around a doctrine and a concrete institutional framework. In turn, the doctrinal differences promote additional factionalizing tendencies with the result that competing churches (as in the case of the Baptists and Presbyterians) become the battleground for a cumulative and circularly reinforcing struggle between opposing factional groups.[14] It is no historical accident that the Baptists in Homestead originally organized as a "protest" against the dominance of the Presbyterian "clan" and that the "clan" retaliated seven months later by re-organizing the Community church, affiliating it with the state Presbyterian board, and taking several additional Homestead families into its power structure.

The many other small denominations within the community do not have churches, but the adherents hold on to their separate beliefs with great resistance and vitality. A particularly crucial case is that of the local "communist and atheist" who publicly announces his deviant beliefs and states as justification for his position that "all I ask of any man is that he make up his own mind instead of takin' somebody else's word fer somethin'." While it might have been assumed that a person with "communist" beliefs would tend to be collateral in orientation, the Cultural Orientations Questionnaire revealed an almost entirely (five responses out of six) individualistic profile for him.[15]

In terms of social rank, the Presbyterian "clan" members regard themselves to be "a notch above" the extended family which controls the Baptist church, and the question can be legitimately raised as to whether these two factions are not two "social classes" in the social structure. This is a complicated problem in a group as small as this where all relationships are highly personalized. If one were to transport the families in Homestead with their present values and attitudes and their material possessions to a larger American town, they would fit approximately into the social status system in the range from Lower Lower to Lower Middle Class (in Warnerian terms).[16] For there are a few individuals at one end of the range who fit the picture of the Lower Lower ("Tobacco Road") status position because of their "shiftless" habits, their "filthy" houses, and their lack of concern for "morals." At the other extreme, there are several families whose houses, cars, and other material possessions and whose interests in higher education, in art and music, in family lineage, in associations with middle-class people in Gallup and Albuquerque make them Middle Class. There is also a difference between the "old settlers" and "newcomers," the former having a more prestigeful position. However, the bulk of Homestead families fall solidly in between these two extremes and could best be described as "good, respectable, working-class people." The Homesteaders describe themselves as "plain old bean farmers," and they do so with considerable pride.

The conceptualization of these two factions as social classes in a stratified structure is not too meaningful in the Homestead situation, for when the total membership in each of the important factions is carefully analyzed it is evident that each faction contains both owners and tenants, persons of low prestige and of high prestige, "old settlers" and "newcomers." Furthermore, some of the Homesteaders with the greatest incomes and the highest prestige are not members of any faction.

The point at which factors of social rank do enter the situation is in the exclusion of the few individuals on the "Tobacco Road" end of the social scale from the more intimate types of social activity. The rest of the families are "acceptable" in "respectable" social circles, and whether they do or do not participate in particular social affairs depends fundamentally upon their relationships to the factional system. Indeed, it might be argued that the *faction system* rather than a *class system* is the "super-organization" in Homestead,[17] for the phenomenon of competing factions within the community both stems from and in turn influences each of the other types of division and as such has more effect upon community action than any of the divisive factors of kinship affiliation, religious attachment, or social rank.

Throughout the history of Homestead, the most stable faction has been the "clan." There has been constant opposition to this large group, but while its core personnel is relatively stable, the most active opposition to it is constantly changing. During the past five years, the "clan" has been most vigorously opposed by the faction which controls the combination store and *café* and the Baptist church. This conflict not only is the most dramatic, but also produces the most important "feuding" from the point of view of the total social system. From these two important factions, the structure shades off to groups of two or three families in smaller factions and finally to individual nuclear families "who only come to town when they run out of spuds and Bull Durham" and "don't associate with anybody very much."

Although the majority of the Homesteaders do not belong to any of the current factions and are quite content to "mind their own business" and stay out of the disputes, the presence of these factions has significant implications for the functioning of the social system. Indeed, if it were not for the initiating and organizing activities of these factions, Homestead would be merely a collection of isolated farmsteads, and there would be little or no "community life." In other words, factionalism can, from a certain point of view, be seen as an organizing principle rather than a disintegrating force in the social system. It is true that the relationship between factions is a conflicting and "feudin'" one, but at least there is a relationship and

THE ATOMISTIC SOCIAL ORDER 169

a higher degree of organization than there would be if every family stayed out on its "place" and only came to town when they ran out of groceries.

The important implication for social action in these factional alignments is that there is a continual power struggle between two or more factions for the control of community affairs. The effects of these conflicts are particularly critical in the organization of the churches, the school system, the baseball club, the rodeo, and the associations such as the Water Coöperative and Farm Bureau. In 1950, the struggle had reached the point where most people avoided being placed in leadership positions for any organized activity, because they immediately became subjected to the most vicious criticism from factions which were not as fully represented in the activity or from those which opposed the faction to which the leader belonged. There is a common saying to the effect that "the best thing to do is to vote your enemies into office," and this maneuvering is observable in action. For the most part, the formal leadership positions in the community are held by "newcomers" who are eager to take part in Homestead's affairs and "haven't learned better yet."

In the school system the factional struggle reaches a particularly acute form in disputes over the position of principal. While two or more of the teaching positions have always been held by members of the large "clan," the position of principal has usually been filled by an outsider. The incoming principal always has the problem of trying to maintain a neutral position with reference to the larger factions, because if he aligns himself with the Presbyterian "clan" and coöperates closely with the teachers who come from it, the opposing factions initiate action against him. If he does not "play ball" with the "clan," he has virtually committed political suicide as far as his tenure in the community is concerned. He therefore usually alienates one faction or another and may become the focus for hostility in the factional struggles; in short, he becomes a "scapegoat" in the total situation. Homestead has had a record of ten different principals in the fourteen years since the high school was established. Seven of these principals left under pressure when they became embroiled in factional disputes.

This process may be illustrated by the case of the latest principal to be "politicked out of the community." This man had previously taught in a nearby community, and before he came to Homestead in the fall of 1949 he had apparently sized up the factional situation. He had been a Baptist previously, but joined the Presbyterian church and was introduced around the village by the Presbyterian missionary. After he came to Homestead, he attempted to participate in both the Community and the Baptist churches. But due to his earlier Baptist training and values, he gradually obtained more sympathetic responses from the Baptist faction and eventually played a significant role in the formal organizing of the Baptist church. This move, combined with his attempts to take over leadership and initiate action in a wide variety of situations, brought increasingly negative responses from the Presbyterian "clan," which seized upon the issue of "incompetence" (there had been other incompetent principals before this one who were not opposed by the "clan") to start a political movement against him. At the end of his second year he was literally "politicked out of the community."

The power struggle has also been an important feature in the organization of the local baseball team through the years. The team was organized in the early 1930's and has always been a center of community interest and participation. Again, there is an annual struggle among members of various factions for positions on the team.

The situation in the case of the Water Coöperative and Farm Bureau is somewhat different in that the initial organization and continuing control have been centered in the large "clan." The response of other factions and other individual families has been to deliberately stay out of these associations and to "talk agin' them." Both are felt by some Homesteaders to be infringements upon individual freedom and independence of action, and we have even heard them described as "communistic," which in the case of the Farm Bureau with its generally conservative farm policies is certainly an amusing misnomer.

Finally, at any organized community event, such as a dance, picnic, or rodeo, the atmosphere is laden with expressed expectations that the coöperative situation may not last and that the organization may disintegrate at any time. At a dance, for example, the seating arrangements and the actual participation in dancing and drinking tend to follow factional lines, and there is the general feeling that the dance may not last beyond midnight by which time the male participants are "likkered up" and the "fights may break out at any

THE ATOMISTIC SOCIAL ORDER 171

minute." This pattern is more marked when compared to the organized sequence of events in a Navaho curing ceremonial, a Mormon dance, a Spanish-American fiesta, or a Pueblo katchina dance, where there is every expectation that the activity will be carried through to a successful completion.

In conclusion, we point out that the feature of the Homestead factionalism which is particularly disintegrative with respect to action at the community level is not the mere presence of such factions (which are probably present in some form in most cultures in the world) but the fact that there is no culturally patterned way in which the resulting power struggles are controlled, channeled, or kept within reasonable bounds. In many nonliterate cultures the factionalizing tendencies become crystallized around lineages, clans, or in more explicit form around moieties which include all members of a given cultural group or community and provide culturally approved ways in which the inevitable conflicts are controlled and channeled in a manner that does not completely disrupt the social organization. Some tendencies in these directions are found in the case of political parties and church groups in Homestead, but these do not extend widely or deeply enough in the community structure to temper the continual factional disputes.

THE PROBLEM OF VARIANCE

In the Introduction, the dominant profile of value-orientations was presented, and in subsequent chapters the relationships between this dominant profile and the on-going social system in Homestead were discussed in detail. It is also clear from the evidence presented in the Appendix that variations exist in the present value system of the community. Many of these variations can be traced to particular experiences in the life histories of individual Homesteaders which have led them to attach higher values to coöperation, or to live more for the present than for the future, or to feel that there is no hope that the natural situation in Homestead will ever be mastered. Over and above these individual factors, however, it is apparent that the culture of Homestead manifests a second-order emphasis upon "collaterality," upon the "present," and upon more harmonious relationships with nature as expressed by the "in-nature" responses on the Cultural Orientations Questionnaire. These more general variant tendencies in the cultural system can, I think, be traced to two fundamental features of the present situation: (*a*) the fact that the rigors of the physical environment have far exceeded the expectations of the original settlers, and (*b*) the fact that Homestead is a marginal community located far from the centers of economic and political power and far from centers of organized religion, etc.

The rigors of the physical environment have generated in several members of the community the attitude that the central values in the dominant profile are hopelessly out of gear with the realities of the situation. In the total population, these people are few in number,[18] but they might be described as the "intellectuals" of the community. They are not "intellectuals" in the sense of having more formal education; some have had only a grade-school education. However, they are characterized by a more thoughtful and flexible approach to their problems as compared to other members of the community.

The factor of marginality is also important in generating and sustaining variance in the value system in that Homestead provides a kind of "haven" where variant persons can "relax" on an isolated farm and escape from the kind of penalties they would face if they lived back in the "Bible Belt" in Texas or in one of the larger centers of Anglo population in New Mexico. It is, for example, possible for the two professed atheists, as well as for many others who are distinctly not active churchgoers, to lead their own lives in Homestead (where active religious participation is given only slight emphasis) and to be free of the pressure for conformity that they would inevitably be subjected to in Plainview, Texas. Other Homesteaders, who prefer to devote their main efforts to a life of hunting and fishing — a life-style emphasizing "present-time" — rather than to a life of farming and hard work, may "get by" on very little by way of food, clothing, and housing and have plenty of time to hunt and fish.

In other words, the loosely structured atomistic social system and the marginal location of Homestead provide a setting in which individuals with variant values may lead satisfying lives while at the same time the larger part of the community adheres to the dominant Homestead profile of value-orientations.[19]

8

Conclusions

The purpose of this final chapter is to set forth the conclusions reached concerning the influence of value-orientations upon the settlement, development, and current transformation of the community of Homestead, and to explore the theoretical implications of these conclusions for the role of value-orientations in cultural processes.

FRONTIER VALUES AND EARLY SETTLEMENT

The settlement of Homestead occurred as an aspect of a larger migratory movement from the South Plains states westward to California. The basic situational factors of drought and depression provided the impetus for this movement out of West Texas and Oklahoma, but some of the migrants did not follow the mass movement on to the West Coast. Rather, they deliberately made the decision to homestead in an isolated region of western New Mexico where the economic risks were great but where there was an opportunity to establish permanent farm homes. In the new settlement on the frontier they were without the social support of a larger group of migrants like those who went on to California and lived in or near more urbanized centers. Moreover, evidence has been presented for the view that the Homesteaders were a group of people in whom the values of the American frontier settlers were especially strong and that they formed a genuine twentieth-century pioneer movement. They were not moving under duress of poverty alone, but were seeking fulfillment of values that were cherished in their way of life.

Of central importance in the new homesteading situation was the fact that the settlers could own farms individually (whereas before, for the most part, they had been tenant farmers, hired men, or laborers) and, in more general terms, could control their own destinies — "be their own bosses." They were also characterized by an orientation to nature which defined the physical environment as something to be controlled and exploited by man for the attainment of his own ends; this provided them with confidence in their efforts to build a dry-farming community in this arid land.

The Homesteaders were also strongly future-time oriented and thus could freely leave their ancestral homes behind and enter an unknown and untried farming territory with unbounded optimism for the future prospects of their community. The conditions of life were severe and uncertain in the early days of settlement, but the Homesteaders' stress upon the future provided an effective dynamic orientation which shifted attention from the memories of the past and the grim realities of the present to their hopes and dreams for the future.

This cluster of strategic value-orientations — emphasis upon "individualism," "mastery over nature," and "future-time" — stimulated the kind of pioneering activity that led to the settlement and early development of the frontier type of community found in Homestead. Indeed, it is doubtful if the community would have survived more than a few years if the Homesteaders had not been a selected group (of the Texas and Oklahoma migrants in the 1930's) with a cultural tradition in which these value-orientations were strongly present.

FRONTIER VALUES IN A CHANGING SITUATION

As one looks at the course of events since the initial settlement of Homestead, the central fact which emerges is the extent to which a particular cultural tradition with its strategic value-orientations has been preserved despite the stimulus for change provided by the shift, through migration, in geographical and cultural setting and by the recent changes in the environmental and economic situation. Homestead is clearly a case in which the situation has changed markedly but the central values have not changed accordingly.

The four hard facts about the present changing situation are: (a) by 1952 it had become clear that the climatic pattern promised only three good farming years in each decade; (b) the continuing practice of planting beans leaves no stubble on the land and the topsoil is gradually blowing away during the spring windstorms; (c) Homestead's total land base has decreased from 130 to approximately 100 sections of land, and the thirty sections which have been lost have

CONCLUSIONS 175

passed into the hands of large ranchers whose holdings now surround the community; and (d) whereas the original holdings of each Homesteader family were one section in size, the average is now two sections and many individual families have much more land.

In response to the original shift of situation from the Plains to New Mexico, the following innovations have appeared in the culture of Homestead: a shift from the farming of cotton and/or wheat to the cultivation of pinto beans as the chief source of livelihood; the picking of pinyon nuts by techniques learned from the Navahos; the use of pinyon logs in the early days of settlement and later the use of adobe bricks (a technique learned from the Spanish-Americans) to build houses; the flourishing of the water-witching pattern; and the development of community dancing — most of the Homesteaders did not know how to dance when they arrived in New Mexico since dancing was strongly frowned upon back in the "Bible Belt."

In response to the more recent changes in the physical environment and economic situation, the Homesteaders have learned to list their fields, plant cross-wind, and practice strip cropping; they have (with enthusiasm stimulated by their belief in mastery over nature) learned to utilize high-powered tractors and associated farm machinery; and those who have managed to acquire the larger landholdings have learned something about the cattle-ranching pattern and are beginning to shift from farming to livestock raising.

But these changes in techniques clearly have not been accompanied by changes of comparable magnitude in the central value-orientations of their cultural tradition. These have been left largely untouched or, in some cases, have actually been intensified by these changes in cultural content. The stress upon "rugged individualism," for example, continues unabated. There has been some tempering of the stress upon "mastery over nature" by the natural vicissitudes of drought and frost, but the exploitation of the land by planting beans continues. The optimistic belief in future progress for the community has hardly been shaken and, in fact, was strongly reemphasized by some of the community leaders in almost nativistic fashion during the drought of 1951–1952. The working-loafing orientation was reinforced by the shift from cotton and wheat farming to bean farming, because the latter permitted full expression to be given to off-season loafing. The stress upon group-superiority vis-à-vis the Spanish-Americans and Indians has undergone a change to the extent that the relationships no longer resemble a strict "caste" system, but as indicated in Chapter 6, the attitudes are far from tolerant and respectful.

THE CURRENT TRANSFORMATION OF THE COMMUNITY

In view of the persistence of the original frontier value-orientations, what is happening to the community of Homestead? What are the positive and negative effects of this *change of situation* but *persistence* (or intensification) *of value-orientations* upon the survival of the community?

With regard to the situation, it is clear that the climatic pattern, the continuing emphasis upon beans instead of other crops, and the encroachment of the large ranchers are all factors which threaten the continued existence of Homestead. In so far as individual Homesteader families acquire more land within the community and are able to shift from straight farming to more ranching, there is a better chance for these families to remain in Homestead; but the effect is also a decline in total population since fewer families can be supported upon the land base by a combination farming-ranching pattern.

The effects of the persisting value-orientations upon the long-range survival of Homestead are seen most clearly in the decisions made by the Homesteaders as they consider the alternative courses of action in (*a*) deciding whether to leave or to remain in the community; (*b*) deciding what economic activities to engage in when they do remain in the community; and (*c*) deciding what action to take in community enterprises.

The prevailing decisions made by the Homesteaders as they respond to these problems in terms of the central value-orientations may now be examined. Hardly a year passes in Homestead when many or all families do not consider the possibility of leaving the community to seek employment elsewhere. Here the evidence is clear, as I have demonstrated in Chapters 2 and 5, that despite more secure economic alternatives elsewhere, most Homesteaders choose to remain in the community and assume the climatic risks rather than abandon the independence of action they cherish and the leisure they enjoy for the more routinized and subordinate roles they would occupy elsewhere. It is significant in this connection that when the

CONCLUSIONS 177

Homesteaders do leave the community to work during drought years, they report they always meet people who ask dozens of questions about Homestead and express deep desires "to have a little place where a man can have his own farm and be his own boss." Homesteaders return to their community reassured that they are enjoying a good life and with the faith that "next year we'll make it." It follows that the Homesteaders are not weighing sheer economic alternatives but are responding in terms of their value-orientations in these decisions; and the effect is one of keeping families in the community, thus contributing to its survival.

On the other hand, when a Homesteader decides to leave the community (and there have been a few almost every year who have finally reached this decision), he often "up and sells out to a big rancher" with little feeling of loyalty to the economic interests of his fellow Homesteaders. This is not merely a matter of the ranchers paying more for the land, for in the case of the land sales during the past five years, there has been at least one Homesteader prepared to make offers had he been given the opportunity before the land was sold to the ranchers. The intra-community individualism and feuding is such that a departing Homesteader feels no real obligation to offer land for sale to his fellow Homesteaders. For example, in the summer of 1950 the rancher living south of Homestead came riding up to the home of one of the bean farmers whose half-section of land adjoined the rancher on the north. The rancher asked the farmer how much he would take for his "place." When the farmer replied, "$3,000," the rancher said he would take it and returned that same afternoon with the papers to close the deal. Again, in the winter of 1950 one of the older Homesteaders decided to sell out and move to Albuquerque. He entered into negotiations with the same rancher (without informing any of his fellow Homesteaders) and sold two sections of land. The following summer one of the younger Homesteaders decided to take a job in El Paso and wanted to sell his land. In this case he did offer the section for sale to one of his neighbors with whom he was on close terms. When his neighbor declined, he sold the section to a rancher without further notice to any of the other residents of Homestead. In all three cases there were other Homesteaders who would have liked to purchase the land but who belonged to other factions in the community and were, consequently, never given the opportunity.

While a few people in Homestead criticize this individualistic free-market behavior, more of them say, "After all, a man ought to be able to do what he wants to." The ultimate effect upon the community, conceived as an "up-and-coming town," is destructive, because the rancher places more cattle upon the section of land and the population of Homestead is reduced by one family.

Each year there is also discussion of possible economic alternatives to bean raising, such as shifting to other crops, to raising livestock on feed grown on the farms, or to raising chickens. Nevertheless, the stress continues to be upon beans, because most farmers are convinced that they constitute the best crop for maximum economic returns. It is evident, however, that this orientation of mastery and exploitation is strictly short-range from the point of view of future survival of the community. Each year there is a drop in the fertility of the bean farming land as the windstorms strip off the topsoil.

The Homesteader families who have been able to shift more than superficially to other economic alternatives have done so by acquiring more land. It is significant that the same Homesteaders who "cuss the ranchers for getting all the land" will make every effort to acquire additional land themselves. This land is added to their original homesteads, and the effect is again a reduction of the population, because new families are not brought in to utilize the land; rather, the additional land is used to raise cattle and to maximize one family's income. While the ranchers have acquired thirty sections that were once part of Homestead's land base, fifty sections have, in the past, been involved in these transactions in which individual families within the community have acquired more land from other Homesteaders. Both types of property transaction result in fewer small family-owned farms in the community each year; thus, whether drought occurs or not, the population steadily declines.

It would appear, then, that both positive and negative effects upon the community's survival arise out of the same value-stress upon individualism. On the one hand, the desire to be one's "own boss" has the effect of keeping families in Homestead. On the other, the free-market behavior results in a double-pronged process of Homesteaders selling out to ranchers (when they do decide to leave Homestead) and of other Homesteaders attempting to maximize their incomes by acquiring more land within the community — a process which leads in both cases to population decline.

CONCLUSIONS 179

The persisting value-orientations are also brought to bear in the Homesteaders' decisions as to what courses of action to follow on various community enterprises which are suggested within the community, or proposed by government officials or others from outside the community. The continuing value-stress upon "individualism" has been most damaging to coöperative efforts such as building the school gymnasium, maintaining organized baseball and basketball teams, managing the annual rodeo, and developing a more effective coöperative for the community well. The Homesteaders are paying a high price for their individualistic values. The stress upon leisure-time loafing also obstructs serious efforts to organize and complete community enterprises. Despite this, the stress upon the future keeps alive the hope that the community will grow and prosper and stimulates the repeated forming of "booster" committees for the promotion of improved roads, a telephone line, etc. The conflicting pressures of these various value-orientations tend to perpetuate a "vicious circle" in which the Homesteaders maintain an "idealized" conception of their community and their hopes for its future expansion, but also continue to stress their orientations of "individualism," "mastery over nature" (but with the emphasis upon short-run economic gains such as the bean cash-crop) and "loafing" in a manner that effectively prevents improving and expanding the community.

THE PERSISTENCE OF VALUE-ORIENTATIONS

The question may now be raised as to why the central value-orientations of the Homesteaders have endured in the face of marked changes in the environmental, economic, and cultural situation, and have persisted to the point where they have become dysfunctional for continuing development (as defined by the "booster" committees) of the local community. One might assume that adjustments would have been made within the community as Homestead entered a declining stage, or that the Homesteaders might have drawn certain values from the neighboring cultural groups with whom they have been in contact during the past twenty years.

The value-orientations which persist in the face of the changing situation in Homestead may well have made functional "sense" at one time. The community was established in an area where according to government officials, ranchers, and old-timers in the Southwest it was impossible to live by dry-farming techniques. In consequence, the Homesteaders feel strongly that the way of life which they have struggled so hard to establish is worth preserving. Their values hang together to provide meaning and purpose to their existence, and there is a deep and widespread feeling that if these values are surrendered, then the whole life-way may collapse. They cling to these values even though there are contradictions between them; and it is apparent to an observer that the meaningfulness is not a matter of logic but a result of a long process of development of a system of values which had its roots in the earlier historical experience of the group and is perpetuated in this frontier community in New Mexico.

It is also evident that value statements differ from existential statements in any given cultural tradition with respect to the readiness with which it is possible to subject them to a testing against the realities of a situation. In the case of an existential statement such as, "It rained two inches yesterday," it is readily possible to obtain conventional agreement as to what does or does not constitute two inches of rain and to check up on this statement. But in the case of value statements concerning what is "desirable" or "nondesirable," the statement of desirability has less ready referent points. Although there is a backward-looking and forward-looking aspect to value-orientations in their function as regulators of action, the regulation is not so much in terms of adjustment to the realities of a situation as it is in terms of the preservation and elaboration of the central values of a way of life once these values have been selected in the course of the history of a given cultural group and are carried along in its cultural traditions.

It must also be noted that the crucial value-orientations of a culture are apparently learned in early childhood, when they are internalized and become aspects of the personality system.[1] Homestead was settled by adults whose values were already deeply embedded in their personality systems. As adults, they could learn how to plant and till pinto beans, how to pick pinyons, how to dance, etc., but to change their values radically would entail a reorganization of their personalities that would be virtually impossible except through some type of traumatic experience. As adult members of the community, it is also clear that they were successful in transmitting their values practically intact to the younger generation which was born and reared in Homestead. It is noteworthy that the

CONCLUSIONS 181

Homesteaders verbalize this socialization process by saying that "you should never go back on your raising."

Finally, it should be noted that relationships with the neighboring cultural groups which could conceivably furnish alternative value systems for adoption by the Homesteaders are of such a nature that significant value transmission is virtually impossible. The Indians and Spanish-Americans are regarded with such contempt by the Homesteaders that they do not take a second look at the possibility of borrowing value-patterns from these groups. With regard to the Mormons, the religious barrier and the "clannish" social attitude of the Mormon community prevent effective transmission of values.

The one neighboring group to which the Homesteaders look with more admiration are the "big ranchers," but in this case the same values cherished by the Homesteaders are found in even more intense form among the ranchers. Thus, from this direction there is reinforcement rather than change for their present values.

The other sources of stimulus for change are those provided by the metropolitan world and those which the ancestral regions of Texas continue to provide. While these sources represent important stimuli for change in cultural content, the basic value-orientations represented by these streams of influence do not vary significantly from those described for Homestead.

It is particularly evident that many Texas Panhandle towns such as Plainview, Lubbock, Hale Center, and Abernathy are still within the social universe of the Homesteaders. Although the Homesteaders say they are now "Texicans" (combination of "Texan" and "New Mexican"), they maintain their contacts by visiting relatives and friends in these Texan communities where they feel very much at home. They also prefer the music of Eddie Arnold and Bob Wills, which is broadcast from the Texan radio stations, to American "popular" or "jazz" music. These contacts with Texan life play an important role in maintaining the dialect, the values, and the identity of the Homesteaders as part of a Texan subcultural group in the modern Southwestern scene.

Furthermore, when the Homesteaders pull up stakes and leave the community for other occupational pursuits in the outside world, they usually move to another location within the Texan (and "Okie") subcultural continuum which now extends from Texas and Oklahoma across the Southwest to California. The emigrants from Homestead generally find people and neighborhoods of their own kind and maintain many of the values they held in Homestead. It is evident that although the community of Homestead as such may be declining, many of the values it represents are being perpetuated elsewhere in the greater Southwest.

CULTURAL PROCESSES

In these empirical materials from our study of the course of events in Homestead there would appear to be two distinguishable types of culture process. One type are the *recurrent* processes which characterize the daily, seasonal, and annual round of life in the community. These are the processes, such as socialization, scapegoat mechanisms, symmetrical schizmogenesis in relations between the sexes, etc., which are often referred to as the "structural dynamics" of a social or cultural system. The emphasis is upon the repetitiveness of the pattern. The second type may be called *directional* processes, which characterize cultural events occurring in functional sequences that are cumulative and circularly reinforcing in their effects over a longer time-span. These are the processes which comprise long-range transformations in a cultural system; elaboration, intensification, crystallization, florescence, growth, drift are all subsumable under this order of process. The emphasis is upon events nonrepetitive within the time-span studied and occurring in a patterned series through time.

The concept of recurrent processes refers to those dynamic features of a cultural system which recur with regularity as individual members are born into and socialized by the social group carrying the cultural system and finally leave the group at death; or recur regularly as the cultural system adjusts each year to the rewards provided by and/or the tensions generated by the natural passage of the seasons; or as the cultural system regularly makes adjustments to contacts with other groups, and so on.

This order of dynamics may be illustrated by what happens in Homestead during the periods of tension in the summer when the bean farmers are anxiously awaiting the summer rains. During this time we have observed a marked increase in the vicious gossip and "character assassinations." The incidence of inter-personal and inter-familial feuding increases markedly, and fist fights at the dances are more likely to occur.

CONCLUSIONS

It is evident that the frustrations generated by the physical environment are relieved by displaced aggression against persons in Homestead, which has a "rational" culture (hence, no witchcraft or ghost patterns to serve as outlets for aggression), but has no effective patterns of coöperative activity to deal with drought in a way that relieves tension and frustration. Consequently, the individualistic social order becomes characterized by even more "feuding and fussing," which normally continue at a high level until the tension is relieved by the arrival of the summer rains, and the Homesteaders start cultivating their crops.

The recurrent processes occur with regularity from year to year, but the temporally connected events that resulted in the settlement, the development, and the current transformation of Homestead form a functional series that can be conceptualized as manifesting a cluster of directional processes.

The events in these processes have formed the following basic sequence:

(a) The combination of drought and depression on the South Plains of Texas and Oklahoma in the early 1930's created a situation in which there were groups of landless people working as farm tenants or laborers who held certain values that were not being fulfilled. This situation provided the original impetus for a mass migration out of the South Plains region.

(b) While most of the migrants followed the main stream on to California, there were a number of families who were attracted by open land and who chose to settle on homesteads in western New Mexico where they confidently expected to get a new start in life and to build permanent farm homes.

(c) In Homestead the new settlers found a setting in which their central values were being fulfilled. Through the years, despite more secure economic alternatives elsewhere, a basic core of Homesteader families has remained in the community, resisting the efforts of the government authorities to transfer them to less "marginal" farming areas, and struggling with the periodic risks of drought and crop failures.

(d) But as the local environmental and economic situation changed, the frontier values which were appropriate and self-correcting for early settlement became inappropriate and self-transforming for the community, and Homestead began to undergo a change from a farming village into a ranching region instead of developing into a metropolis.

The characteristics of certain of these directional processes comprise elements that might be conceptualized by Goldenweiser's idea of "involution," by Kroeber's concept of "pattern saturation," and by Cannon's concept of "homeostasis." But the total process consists of something more than the phenomena delineated by each of these concepts, or by all of them in combination.

In describing his concept of "involution," Goldenweiser writes:

> The application of the pattern concept to a cultural feature in the process of development provides, I think, a way of explaining one peculiarity of primitive cultures. The primary effect of pattern is, of course, to check development, or at least to limit it. As soon as the pattern form is reached further change is inhibited by the tenacity of the pattern. While characteristic of all things cultural, especially in primitiveness, this aspect of pattern is particularly conspicuous in rituals and the forms of religious objects, where the tenacity of pattern is enhanced by social inertia or a sacred halo. But there are also other instances where pattern merely sets a limit, a frame, as it were, within which further change is permitted if not invited. Take, for instance, the decorative art of the Maori . . . The pattern precludes the use of another unit or of other units, but it is not inimical to play with the unit or units. The inevitable result is progressive complication, variety within uniformity, virtuosity within monotony. This is *involution*.
>
> A parallel instance, in later periods of history, is provided by what is called ornateness in art, as in the late Gothic. The basic forms of the art have reached finality, the structural features are fixed beyond variation, inventive originality is exhausted. Still, development goes on. Being hemmed in on all sides by crystallized pattern, it takes the function of elaboration. Expansive creativeness having dried up at the source, a special kind of virtuosity takes its place, a sort of technical hairsplitting. No longer capable of genuine procreation, art here, like a seedless orange, breeds within itself, crowding its inner structure with the pale specters of unborn generations.[2]

In his *Configurations of Culture Growth* Kroeber develops the closely related idea of "pattern saturation." He writes:

> As they [the high-value culture patterns] begin to select, early in their formation, they commit themselves to certain specializations, and exclude others. If this arouses conflict with other parts of the culture in which the pattern is forming, the selection and exclusion may be abandoned, the pattern as something well differentiated be renounced, and nothing of much cultural value eventuate. If, however, this does not happen, but the

CONCLUSIONS 185

other patterns of the culture reinforce the growing one, or at least do not conflict with it, the pattern in question tends to develop cumulatively in the direction in which it first differentiated, by a sort of momentum. Finally, either a conflict with the rest of its culture arises and puts an end to the pattern, or it explores and traverses the new opportunities lying in its selective path, until less and less of these remain, and at last none. The pattern can be said to have fulfilled itself when its opportunities or possibilities have been exhausted.[3]

From a certain point of view it seems evident that the present cultural state of Homestead could be viewed as manifesting a condition of "involution" and "pattern saturation." There is evident a tenacity and crystallization in its value system, which appears to have run its course in exploring and utilizing the possibilities which lay in its selective path. The structural features of the value system seem fixed beyond basic variation, and the minor changes taking place from year to year comprise small elaborations within the limits of these basic patterns. The expansive and creative features which characterized the early days of settlement have given way to a monotonous recital of a set of fixed value-orientations. But what must be added to this conceptualization is the dynamic interplay between this crystallized value system and the changing environmental and economic situation, a dynamism which is transforming the community in a direction unforeseen by the original founders.

To understand this dynamic interplay, the concept of "homeostasis" may also be examined as describing another aspect of the sequence of events in Homestead. This concept as developed by Cannon was intended to have broad meaning and implications. After defining homeostasis as the maintenance in the animal body of steady states, by coördinated physiological processes, Cannon goes on to state:

> It seems not impossible that the means employed by the more highly evolved animals for preserving uniform and stable their internal economy (i.e. for preserving homeostasis) may present some general principles for the establishment, regulation, and control of steady states, that would be suggestive for other kinds of organization — even social and industrial — which suffer from distressing perturbations.[4]

A related theoretical development in recent years has been the efforts of Wiener and others to interpret processual phenomena in terms of cybernetic concepts such as "feedback," "automatic control," etc.[5]

It is evident this cluster of concepts has utility for the description of the sequence of events in Homestead. I have pointed out how the original value-orientations of the Homesteaders were reinforced by their early experience in New Mexico which led to fulfillment of these values and for a period of time resulted in community development and expansion. There was an important "feedback" here between the new setting and the Homesteader value system.

As time went on, however, the system was not self-correcting from the point of view of further development and expansion, but rather became self-transforming in a manner that led to preservation, and indeed crystallization, of the value system but at the same time to decline and deterioration of the local community. Richards has recently introduced the concept of "hyperexis" to correct the over-emphasis upon "homeostasis" in our biological and medical thinking. He writes:

> . . . an important category in pathology is the excess response, a homeostatic effort that overreaches itself, to the detriment or even the death of the organism. The term hyperexis is suggested for this type of excess response; a term that appears in Plato's *Timaeus*, and means "having too much." [6]

This concept of "hyperexis" is highly suggestive for delineating the course of events I describe in the long-range directional processes in Homestead. For it would appear that the earlier "feedback" from the new setting to the Homesteaders' value system probably led to an overstressing (a homeostatic effort that overreached itself) of the central values in a manner that resulted in "involution" and "saturation" on the value-pattern level but in "hyperexis" in the dynamic interplay between value system and ecological-economic situation. The frontier values that were appropriate and self-correcting in the early part of the sequence became inappropriate and self-transforming for the community in the latter part.

How common this type of directional sequence is in the growth and decline of cultures and communities is a matter for further empirical research, but it does appear to have been of critical importance in the Homestead case.

THE ROLE OF VALUE-ORIENTATIONS IN CULTURE PROCESSES

Since these directional processes in a cultural system cannot be reduced to and explained in terms of "rational" responses to the

CONCLUSIONS 187

realities of environmental and economic situations, or as responses to biological impulses, let us explore how value-orientations in their functions which I have conceptualized as "selection," "regulation," and "goal-discrimination" influence the outcome of these series of connected events.

Growing out of the historical experience of a given cultural group are certain critical value-choices which represent a selection among a range of possible alternatives. The historical roots of the frontier value system described for the Homesteaders are to be found in the early pioneer experience of the Jacksonian West (with the ceaseless change, vast opportunity, and fluid social structure which characterized the frontier movement in America), combined with a Southern historical experience which provided the sources for their group-superiority and loafing values.

At the time the Homesteaders arrived in New Mexico, certain fundamental preferences as to what is desirable in a way of life were already well developed within their cultural tradition, and the community was composed primarily of strongly individualistic, future-oriented persons who sought to subdue and exploit the natural world.

In their function as regulators, or a kind of cultural "automatic pilot," these orientations have had the effect of keeping social action on the same track previously selected, rather than of adjusting the cultural system to the objective realities of the changing environmental and economic situation. These effects are the consequences of the fact that (a) the values provide a set of cultural lenses through which the Homesteaders view the objective situation;[7] and (b) the values provide a continuing definition of the limits of permissible behavior for the crucial roles within the social order, and transgressions are handled by the social mechanisms of gossip, social ostracism, fist-fights, etc.

Finally, it is clear that the values provide a set of criteria for the ranking and discrimination of goals for future action, which continue to influence behavior from day to day as the Homesteader weighs the alternatives that face him. For example, in 1950 it was still apparent that the value-goal of independence of action took precedence over economic security, and most Homesteaders were choosing to remain in the community — even after the crops were damaged by a severe hailstorm in August and an early frost in

September. It was also apparent that the stress upon individualism took precedence over the value-goal of coöperative activity as nothing further had been done to complete the high school gymnasium, to extend the activities of the Homestead Coöperative, or to gravel the village streets.

The conclusion one must reach is that precisely because the Homesteaders in their firm attachment to their values do not take cognizance of the ecological and economic situation, the dreams of the community's founders will not be fulfilled and Homestead will never "become a city like Plainview, Texas." Instead of changing from a farming village into a metropolis as the founders envisioned (and as the younger generation still hopes), the community is in the process of becoming a settlement of widely scattered ranches. This process is likely to continue to the point where the community will be too small to support its service center of schools, churches, and stores. Indeed, there is the possibility that the community as such will disappear altogether as the ranchers continue to buy up the land, and eventually the "big ranchers" may graze cattle in the streets of Homestead — a frontier community which started out with such big dreams of its future.

APPENDIX

APPENDIX

SUMMARY OF DATA ON THE HOMESTEADERS' VALUE-ORIENTATIONS

The tabulated data in this appendix take the following form. There is first the identification of the *major value-orientations*, which are conceptualized as general value-statements summarizing the patterned relationships which the Homesteaders strive to achieve and maintain with key aspects of their total life situation: relations to other men, to nature, to time, to work (i.e., the allocation of activities from day to day), and to other cultural groups.[1] The statement of a major value-orientation is followed by a presentation of the cluster of *associated values* which are very explicitly and concretely expressed by the Homesteaders. These associated values are presented in the form of *dictates* or *prescriptions* as to the most desirable course of action to follow in dealing with the key aspects of human life as listed above.

The empirical evidence for these values was collected in three forms. First, a series of intensive interviews with twenty selected informants was conducted during 1949–1950. The group of twenty included fifteen men and five women, selected after four months of field work in Homestead to represent a spread of strategic positions in the economic and social structure. For this series of interviews a *random* sample was not drawn because in a community as small as Homestead, it would probably fail to be as representative of the total community structure as a carefully selected strategic sample which included persons of high, middle, and low income, both farmers and business people, both leaders and followers, and persons belonging to different political parties, religious denominations, educational levels, and age levels. These formal interviews were conducted by means of an interview schedule which served as a check list to remind me of important areas to cover but which also permitted long periods of nondirective interviewing. Full longhand notes were taken on these interviews which I either typed myself or recorded on an audograph for later typing soon after the interview was completed.[2]

The second form employed to collect empirical data on value-orientations was the Cultural Orientations Questionnaire, which was administered in Homestead in 1951 to a random sample of twenty informants (ten men and ten women).[3]

Finally, a summary was made of the *observations of recurring situations* (recorded in my field notes or in those of the other project researchers who have done field work in the community) which occurred during the course of our field work or were described by several informants as having happened earlier in the history of Homestead.

In the following tabular summary of these three bodies of data I have presented not only the *dominant responses* but also the *variant responses*, showing that while the value-orientation profile described in the chapters of this book is the dominant one, Homestead, like any other community, has variations within its value system.[4]

I. Major Value-Orientation: INDIVIDUALISM

1. Associated Value: "HAVE YOUR OWN FARM."

Dominant Responses

INT:[*] This value was given concrete expression by all 20 informants.

COQ: Eighteen out of 20 informants chose the "individualistic" responses on the *Property* and *Livestock Inheritance* questions.

ORS: (a) During our year in Homestead each of the 8 tenant farmers made attempts to purchase land. One finally succeeded in 1950.

(b) Homesteaders repeatedly express reluctance to mortgage their land and take great pride in self-ownership.

Variant Responses

COQ: Two informants chose the "collateral" responses on the *Property* and *Livestock Inheritance* questions.

2. Associated Value: "BE YOUR OWN BOSS." ("BE INDEPENDENT.")

Dominant Responses

INT: This value was given concrete expression by all 20 informants.

COQ: "Individualistic" responses were chosen by 19 informants on the *Help in Case of Failure* questions; by 16 informants on *Family Work*; and by 15 on *Employee Relations*.

ORS: (a) When Homestead was first settled, there were enormous difficulties involved in establishing the school because "everybody had a different idea of where the school should be located, and everybody wanted to drive the school bus."

(b) Most Homesteaders find it extremely difficult to tolerate "too many kinfolks" who curb the freedom of the nuclear family.

(c) The leaders in Homestead who are "independent" and lead "by example" have more influence than the "organizers" who are considered "meddlers" in other people's business.

(d) Community gatherings are usually sparsely attended (with the exception of dances) and are marked by an expectation that the interpersonal tensions may reach a point at which the gathering will dissolve. This situation is particularly marked in meetings of the Rodeo Committee, Baseball Club, Farm Bureau and school affairs. A pointed illustration was provided by the course of events during the planning of a camping trip

---

[*] INT refers to the intensive interviews with twenty informants; COQ refers to the Cultural Orientations Questionnaire administered to twenty informants; ORS refers to "observations of recurring situations."

APPENDIX

we made with three Homestead families in 1952. After the departure date had been set, we received word from a Harvard colleague that he was arriving to visit us in Homestead on the day we had planned to leave. We explained to the other families that we had to wait a day and would join them later. The immediate reaction was to jump to the conclusion that we were "backing out" and that we would not show up at all. We found it necessary to reassure the other families again and again that we *would* join them the next day. Sharp comments were also made to the effect that we were letting our Harvard colleague "ride over us" — i.e., that we were not behaving very independently!

(e) Large community projects were failures during 1949–1950 because the Homesteaders failed to coöperate in these enterprises. In the fall of 1949 a suggestion was made that the Homesteaders pool their resources and hire a construction company, which was building a road near Homestead, to gravel the village streets. This community plan was rejected, and an alternative plan was followed. The operators of the various business establishments each independently hired the construction company truck drivers to haul a few loads of gravel to be placed in front of their own place of business, which still left the rest of the village streets a sea of mud in rainy weather. In the spring of 1950, funds for the materials for the construction of a high school gymnasium were offered to the community by the district school board, providing that the residents of the community would contribute their labor gratis for the building of the gymnasium. This plan was also rejected by the residents of Homestead, and there were long speeches to the effect that: "I've got to look after my own farm and my own family first; I can't be up here in town building a gymnasium." By way of contrast, both projects were also presented to the Mormon community of Rimrock, and were accepted by the Mormons. The streets have already been graveled and the gymnasium is nearing completion.

(f) When Homesteaders have left the community for various reasons (drought, health, etc.) in recent years they have sold their farms to the large ranchers whose holdings border on the Homestead community. A few Homesteaders always object to these transactions, but most people feel that it is acceptable behavior and say "a person should be able to do what he wants to."

(g) The Homestead Water Coöperative which manages the community well has repeatedly refused to let individuals pipe water into their houses until everyone has a separate meter and the water can be carefully measured out to each individual family. By way of contrast, the community water organization in the Mormon village of Rimrock has piped water into each house and each family is charged a flat rate per month.

(h) In the Rio Grande Valley, in an area settled by families who previously lived in Homestead and other Homesteader communities of New Mexico, each farmer has been given a separate meter to measure the amount of irrigation water that he uses on his farm. This arrangement became necessary as Texans replaced Spanish-Americans on the irrigated farms in the valley.

(i) On hunting trips, the Homesteaders often want to come back home after they have killed their own deer; whereas Mormon hunting parties are more likely to stay in the mountains until deer are killed for each member of the hunting party.

(j) The Spanish-Americans in Tapala repeatedly comment that Homestead has no *patron* and say that among the Texans "everybody's a big shot."

(k) In negotiations with the PMA (and other government agencies) the Homesteaders usually feel that you can't trust the government and that you must rely on your own efforts in the long run to get things done.

(l) Many Homesteaders criticize the Farm Bureau for being "communistic and socialistic," whereas in point of fact the Farm Bureau is a very conservative farm organization.

(m) In community affairs there is constant "buck-passing" and shirking of responsibility.

(n) Large family gatherings on Thanksgiving and Christmas (which are an old American pattern) are the exception rather than the rule in Homestead.

(o) There is a pattern of early independence training for children in Homestead. Children are expected to be self-reliant at an early age — to learn to drive cars and tractors, to haul wood and water, to work in the fields, to handle money, etc. (By the age of 10 or 11 most Homestead children can do all of these things.) There is little interest on the part of adults in children's play activities. The children are constantly told to go outside or in other rooms to play by themselves. (The pattern is completely different from the "child-centered" urban middle class families.) Children are also often given the crop from part of the field and are expected to work it themselves and to spend the money after the crop is sold. In doing wagework for others, teen-age boys are typically paid adult wages. In general, children show a great desire to be grown up and "on their own."

(p) When I moved to Homestead, I made an effort to do favors for the Homesteaders by making the rounds of the business establishments and asking them if they wanted me to bring anything when I made a trip to the city. This practice evoked reactions of surprise because most Homesteaders make trips to the city without doing errands for one another.

(q) We found it impossible to find people to haul wood and do laundry for us when we first moved to Homestead. It is assumed that every self-respecting man will haul his own wood and that his wife will do her own laundry.

(r) The Homesteaders repeatedly expressed surprise that the research directors could tell other field workers what to do. By contrast, the Spanish-Americans assumed that the research enterprise must have a single *patron*.

APPENDIX 195

(s) Whenever one person does a favor for another in Homestead, there is an almost compulsive tendency to repay this favor as soon as possible, whether the favor be an invitation to a meal, a gift, or a day's labor. An illuminating example was provided by the fact that we received a 66 per cent return on Christmas cards we sent out to Homestead residents in 1950 — as compared to a 25 per cent return from the cards we sent to the residents in Rimrock. When we explored the reasons for this pattern of "paying back favors," we discovered that the Homesteaders do not like to be dependent upon, or in debt to, anyone for favors done.

(t) All Homesteaders forced by drought to leave the community for wagework in the winter of 1950 returned "to try it again" in the spring of 1951. All expressed the desire to have the freedom and independence of action enjoyed in Homestead — despite less opportunity for economic security in the community.

Variant Responses

COQ: On the *Family Work* question four informants chose the "collateral" response; on *Employee Relations* four informants chose the "collateral" and one chose the "lineal" responses; one informant chose the "collateral" response on *Help in Case of Failure*.

ORS: (a) In certain critical areas of community life there are limited patterns of coöperation. The only formally organized coöperative is the one which manages the community well, and even this coöperative does not work smoothly. In the first place, it has never had more than a fourth of the families in Homestead as members. There have also been continuing difficulties in the management of the well and the sale of water. For many years the system was to "sign up" on a sheet of paper posted at the well, whenever one hauled a barrel of water. But many people "failed" to "sign up"; others refused to pay their bills at the end of the month. Finally, in 1950, a coin-operated electric switch was installed and they now have to put a dime in a meter which measures out 55 gallons of water. Furthermore, efforts to expand the activities of the coöperative have repeatedly failed. "We've sold pressure cookers and posthole diggers, but that's as far as we've ever gone," reported one of the old-timers.

(b) Throughout the history of the community there has existed a pattern of "swap work" or "change work," but the actual operation has always involved much grumbling and shifting alliances. In the early days of the threshing machine there was particular need for this swap work during harvest time, but the advent of the combine has made the pattern less necessary. However, the pattern still exists among the women in the care of one another's children when the women make trips to the city.

(c) The organization of the school's hot-lunch program in 1949 was characterized by good coöperation on the part of most families in the community.

(d) Major crisis situations evoke patterns of community-wide coöperation. These are particularly evident in the case of deaths and funerals, during which time almost every family comes forth to offer its services and to attend the funerals. The burning down of a farm house was also a crisis of sufficient magnitude to result in community-wide coöperation, which included both the attempt to put out the fire and a benefit party for the unfortunate family.

3. Associated Value: "EACH MAN SHOULD HAVE AN EQUAL VOICE IN COMMUNITY AFFAIRS." (DECISIONS SHOULD BE REACHED BY POPULAR VOTE.)

Dominant Responses
INT: This value was given concrete expression by all 20 informants.
COQ: The "individualistic" response was chosen by 13 informants on the *Well Arrangements* question and by 10 informants on the *Delegate* question.
ORS: (a) The selection of persons for formal positions on committees or in other organizations in Homestead is always done by voting; each adult present has one vote and the group abides by the decision of the majority. The procedure is carried out painstakingly even when only a handful of people are present at a meeting. By way of contrast, it should be noted that in the Spanish-American community such positions are filled by tradition or by appointment.

Variant Responses
COQ: Ten informants chose the "collateral" response on the question of selecting a *Delegate*. It should be noted, however, that the "collateral" procedure (in which there is a general discussion until almost all of the group agrees — the method which was followed aboriginally by the Navaho, for example) has never actually been observed in Homestead. The "collateral" responses were probably the result of a widespread feeling in Homestead that the "meddlers" are organizing things and getting elected to office and that it would be preferable to have more discussion at meetings. Seven responses were "collateral" on the *Well Arrangements* question.

4. Associated Value: "KEEP UP WITH YOUR NEIGHBORS." (INDIVIDUAL COMPETITION)

Dominant Responses
INT: This value was given concrete expression by all 20 informants.
COQ: There were no relevant questions on this value.
ORS: (a) During our year in the community there was constant competition among individuals in the production of larger bean crops, in the acquisition of more material goods (automobiles, tractors, etc.), and in the building and improving of houses.
(b) Even within related family units, the competition was evident in the rush to purchase new deep freezes, coffee percolators, washing machines, etc., when the power line reached Homestead. In this competitive

APPENDIX 197

buying there was also a great deal of envy openly expressed of persons who had more material goods.

(c) During the meetings of the bridge club, at which the writer's wife was asked to serve as the instructor, more interest was shown in the scoring system than in the fundamentals of the game. Each member of the club had to know exactly where she stood on the scoring!

Variant Responses: None

5. Associated Value: "BE FREE OF THE LAW."

Dominant Responses

INT: This value was given concrete expression by 15 of the 20 informants interviewed.

COQ: There were no relevant questions on this value.

ORS: Although the Homesteaders conform rigidly to marriage and divorce laws and to property laws, they resent further interference from any outside authority, which is called "the law."

(a) The Homesteaders are not anxious to assume too much authority, but they do not want others to, either. One of the major "gripes" against the school principal during 1950–1951 was that he assumed too much authority and "was always trying to tell others what to do."

(b) For years the sheriff's office at the county seat tried to deputize somebody in the community, but nobody wanted to be deputy sheriff "because people get it in for you if you start making arrests." One of the "newcomers" was finally made deputy sheriff, but he has never made an arrest. His attempts to be helpful in crisis situations have always been rejected. In general, the Homesteaders laugh at the deputy sheriff and say that "he would go to bed" if faced with a difficult case.

(c) In 1947 one of the Homesteaders committed suicide (or was murdered, depending on which faction in the community is telling the story). Again, the situation was handled locally, including the inquest and burial. The state police were not called in on the case.

(d) The Homesteaders thought it was terrible when, in 1948, one of the newcomers in Homestead, who ran a drug store, reported to "the law" that some teen-age boys were shooting firecrackers around her drug store. It was strongly felt that the situation could have been handled locally.

(e) The hunting of deer takes place year around in Homestead, and even the game warden participates. The game regulations are regarded as laws which save the deer for city people. It is felt that the deer belong to the community and can be hunted at any time for meat.

(f) The local shop which issues automobile brake and light stickers has never been known to give the required test.

(g) The Homesteaders freely use farming gasoline which is tax-free in their passenger cars. When the inspector arrives in the community, the cars using tax-free gasoline are hidden in the woods.

(h) The local store sells oleomargarine without a license. (It might be added that the storekeeper is a pillar of the Baptist church.)

(i) During the time the Veterans Administration class was in session in the community, most of the veterans had jobs on a road construction project despite firm VA regulations that a veteran should not take an outside job. This fact was well known and approved of by the community, and there was deep resentment when "somebody turned the GI's in." The VA supervisor who came around periodically to inspect the class was called the "snoopervisor!"

(j) The Homesteaders have been eager to receive benefits from the PMA and other governmental agencies, but they deeply resent the regulations of these agencies which curb their individual freedom as farm operators. As discussed in Chapter 4, they rejected the proposals of the Federal Security Administration to move them to the Rio Grande Valley on irrigated land; they have also taken a dim view in recent years of the bean acreage allotments issued by the PMA office.

(k) The anti-authoritarian position of the Homesteaders was also evident when a dance was given in 1949 as a response to the Baptist minister's revival meeting sermon against dancing.

Variant Responses
INT: Five of the 20 informants interviewed felt that "the law" should be called in more often to deal with community affairs.

II. Major Value-Orientation: MASTERY OVER NATURE

1. Associated Value: "USE MACHINES AND OTHER MODERN METHODS TO EXPLOIT THE ENVIRONMENT AND MAKE FARMING MORE EFFICIENT."

Dominant Responses
INT: This value was given concrete expression by all of the 20 informants interviewed.

COQ: The "mastery" responses were chosen by 16 informants on the *Livestock Dying* question, by 18 informants on *Use of Fields*.

ORS: (a) The Homesteaders are constantly striving to use new and more powerful machines in their farming operations, and Homestead is now the most highly mechanized community in the area.

(b) Bulldozers are used to clear the land; tractors are used to pull farm implements; tractor-driven post-hole diggers are used in building fences; the small-grain combine has been adapted to bean threshing.

(c) The "bean knife-sled" (modeled after older and more cumbersome types used in Texas) was developed in Homestead as the most efficient method of cutting weeds early in the summer and for cutting beans during the harvest.

(d) Homestead is the only cultural group in the area in which the number of tractors exceeds the number of horses owned by members of the community.

(e) Farmers who do not have the latest and most efficient farming equipment have less prestige and are subjects for "talk" and criticism.

APPENDIX   199

(f) Every Homestead family owns at least one automobile, and everyone would like a new car.

(g) The arrival of the Rural Electrification Administration power line was greeted with enthusiasm, and all houses in Homestead within reach of the line have electricity. By way of contrast, only one house in the Spanish-American village has been wired for electricity.

(h) The Homesteaders maintain a standing committee to promote the installation of a telephone line for the community. Meanwhile one family has installed a "barbed-wire telephone" which uses the wire on fences to connect three neighboring farm houses.

(i) The Homesteaders manifest great interest in television, and the suggestion has been made that a TV station will eventually be built on top of a nearby volcanic crater in order to reach all the homes in the area.

(j) In the building of homes the Homesteaders show great interest in having the "most modern" conveniences and appliances. This includes fluorescent lighting in living rooms and fireplaces with butane heaters.

(k) The Homesteaders use up-to-date methods in doctoring livestock in cases of illness in the herds and summon veterinarians in serious cases — or at least get the advice of professional veterinarians when it is impossible to get one to come out from the city.

(l) The Homesteaders try to improve their herds of cattle by breeding with registered bulls when possible.

(m) The Homesteaders exploit the land by planting beans as the major crop year after year; this is short-run efficiency because beans bring more money, but a long-range loss to the land, since beans leave no stubble and the topsoil blows away during the spring windstorms.

Variant Responses

COQ: Three informants chose the "subjugated" response on the *Livestock Dying* question; one informant chose the "in-nature" response on *Use of Fields*; 16 out of 19 informants chose the "in-nature" response, one chose the "mastery" response, and two chose the "subjugated" response on the *Livestock* questions.[*] It is probable that the "in-nature" responses on the livestock question (which were markedly different from the responses on the other questions involving the relationship with nature) were attempts to verbalize "good" grazing practices which have been promoted by the Soil Conservation Service and other governmental agencies.

2. Associated Value: "TRY TO CONTROL NATURAL CONDITIONS."

Dominant Responses

INT: This value was given concrete expression by 16 of the 20 informants interviewed.

COQ: The "mastery" responses were chosen by 10 informants on the question on *Facing Conditions*; by 10 informants on *Belief in Control*.

[*] There were only 19 complete protocols of the questions on relationships to nature.

ORS: (a) The Homesteaders practice strip-cropping in an attempt to control the loss of topsoil in heavy windstorms. Bitterness is expressed about the windstorms blowing away the topsoil. "There is always some joking about it, but the smiles don't go very deep," reported one of the old-timers.

(b) The majority of the Homesteaders have expressed great interest in artificial rain-making, and two of them (with the largest landholdings) participated in the rain-making venture during 1951.

(c) Recent attempts have been made to drill irrigation wells and thereby control problems of drought.

(d) Water witching, the use of the phases of the moon, the use of "signs" of the zodiac, and other folk practices are considered (by their adherents in Homestead) as further means of controlling the natural conditions.

(e) When the Homesteaders take camping trips (for example, when they go deer hunting), they usually take along elaborate gear and equipment, such as tents, stoves, and sometimes even chairs and tables.

Variant Responses

INT: Four informants expressed the point of view that it was impossible to control natural conditions and felt that "you just have to take it as it comes."

COQ: On the question of *Facing Conditions*, four informants chose the "subjugated" response and five informants chose the "in-nature" response; on *Belief in Control*, five informants chose the "subjugated" response and four chose the "in-nature" response.

    3. Associated Value: "TRY TO CONTROL HUMAN LIFE."

Dominant Responses

INT: This value was given concrete expression by all of the 20 informants interviewed.

COQ: The "mastery" response was chosen by 13 informants on the *Long Life* question.

ORS: (a) Modern methods of birth control are approved of and practiced by almost all of the younger generation of parents in Homestead.

(b) For illness almost all Homesteaders go to hospitals and utilize modern medical practices.

(c) There is also general approval of and use of vaccinations to prevent illness.

Variant Responses

COQ: On the *Long Life* question two informants chose the "subjugated" response and four informants chose the "in-nature" response.

    4. Associated Value: "WE DO ALL WE CAN AND AFTER THAT IT'S A MATTER OF LUCK."

Dominant Responses

INT: This value was given concrete expression by all 20 informants.

APPENDIX

COQ: There were no questions on the questionnaire that were directly relevant to this value.

ORS: (a) The Homesteaders speak of farming as a "gamble" when there is a crop failure. They also say that "you hit" when you make a crop, and you "don't hit" when you fail to make a crop. (The responses are comparable to "hitting a jackpot" on a slot machine.)

(b) One of the most frequent jokes in Homestead is the definition of a bean-farmer — "a bean-farmer is a man who is crazy enough to think he is going to make it next year."

Variant Responses: None

III. Major Value-Orientation: LIVING FOR THE FUTURE

1. Associated Value: "BE PROGRESSIVE."

Dominant Responses

INT: This value was given concrete expression by 16 of the 20 informants interviewed.

COQ: The "future" response was chosen by 9 informants on the *Child Training* question; by 18 informants on the *War Plant* question; by 9 informants on the *Ceremonial Change* question; and by 10 informants on the *Water Division* question.

ORS: (a) As soon as Homestead was settled, the Homesteaders began to make elaborate plans for the growth of a town at the crossroads in the center of the community. An informal weekly newspaper was written by two of the early settlers and copies were sent back to Texas and Oklahoma to induce more settlers to come to Homestead.

(b) Later this "boosterism" took the form of an organized "Booster's Club" and a weekly newspaper called *The Homestead Promoter*.

(c) There have always been countless committees in Homestead whose functions have been to foster "progress" — road committees, school committees, telephone committees, etc.

(d) The Homesteaders have always felt that Homestead would grow into a modern community with paved streets, movie theaters, department stores, etc., if they could keep up the "booster spirit."

(e) The Homesteaders have always looked forward to the time when the Democratic "Anglos" would control the county politics instead of the Republican Spanish-Americans and have felt that this change will represent "real progress" in good government.

(f) The Homesteaders point with pride to the material progress that the community has made in their lifetime. They used to use horses and now they use tractors. They first threshed beans by hand; then they used the old-fashioned threshing machine; now they use up-to-date combines. They first used kerosene lamps, later gasoline lamps, now they have electric lights.

(g) There are "big plans" for "modern" housing in Homestead. The

Homesteaders pour over house plans in *Better Homes and Gardens* in thinking about their future homes; the local plumber expects eventually to install bathrooms in every house in Homestead; and every family now looks forward to having a full set of electrical appliances in its home.

(h) The Homesteaders who still live in the original log cabins are sensitive and apologetic about this fact. When the observer started to take a photograph of one of these cabins, the farmer commented (apologetically), "I guess everything around here is old," and requested that he be photographed beside his new Chevrolet automobile.

(i) The Homesteaders criticize the Spanish-Americans and Indians for not being "progressive" and point out that these groups have not made many changes in their ways of living in the past 20 years.

(j) The most frequent ethnic jokes told about Spanish-Americans and Indians involve their lack of progressiveness and their lack of sense of working for the future.

Variant Responses

INT: Four informants expressed the point of view that things were pretty good as they were and there was no reason to try to improve the community.

COQ: On the *Child Training* question 8 informants chose the "present" and two informants chose the "past" responses; on the *War Plant* question one informant chose the "present" response; on the *Ceremonial Change* question 5 informants chose the "present" and 4 informants chose the "past" responses; on the *Water Division* question 5 informants chose the "present" and 3 informants chose the "past" responses; on the *Philosophy of Life* question 11 informants chose the "present" response as compared to only 8 informants who chose the "future" response.*

2. Associated Value: "BE OPTIMISTIC."

Dominant Responses

INT: This value was given concrete expression by 16 of 20 informants interviewed.

COQ: The "future" response was chosen by 15 informants on the *Expectations* question.

ORS: (a) Homestead was popularly called "The Garden of Eden" by many of the original settlers.

(b) The Homesteaders attempt to maintain real estate lots and other property near the center of the community so that they can profit "when the four-lane highway is built through Homestead."

(c) The "squatter" who has refused to be moved off the land by the Bureau of Land Management has for 16 years optimistically looked forward to obtaining the right to homestead.

* There were only 18 complete protocols of the questions on *Ceremonial Change* and *Water Division*, and 19 complete on the *Child Training, War Plant,* and *Philososphy of Life* questions.

APPENDIX

(d) The building of new homes and the modernization of older structures went on even in the face of the severe drought of 1950–1951.

(e) The bankers in the nearby cities described the Homesteaders as "the greatest next-year people in the country."

(f) During the 1950–1951 drought, there was still general talk in the community about how the roads would be paved, the gymnasium would be completed, the people who were moving away would return, etc.

Variant Responses

INT: Four informants were thoroughly pessimistic about their own future and about the future prospects of Homestead.

COQ: Three informants chose the "present" and one informant chose the "past" responses on the *Expectations* question.

3. Associated Value: "TRY TO BE SUCCESSFUL."

Dominant Responses

INT: This value was given concrete expression by 15 of the 20 informants interviewed.

COQ: There were no questions on the questionnaire that were directly relevant to this value.

ORS: (a) Looking forward to and striving for success in life is indicated by the rewards of praise and prestige for those farmers in Homestead who possess larger landholdings and grow larger bean crops, and for Homestead residents who leave the community and become successful in urban jobs.

(b) Conversely, farmers in Homestead who are consistent "failures" are criticized for their lack of industriousness; Homesteaders who leave the community for common labor jobs are "talked about" for a period and eventually forgotten.

Variant Responses

INT: Five informants expressed the point of view that all one could expect in life is "to jest git by from year to year."

IV. Major Value-Orientation: WORKING–LOAFING

1. Associated Value: "WORK HARD WHEN THERE IS WORK TO DO," BUT "LOAF HARD WHEN THERE IS NO WORK TO DO."

Dominant Responses

INT: These values were given concrete expression by 18 of the 20 informants interviewed.

COQ: The evidence provided by the questionnaire on these values is most apparent in the consistent split in the responses between "doing" and "being." In no other category of questions was there a comparable consistent split in results.

Question	Doing	Being
Job choice (employee)	11	9
Job choice (boss)	12	8
Things go wrong	11	9
Joy vs. deeds	12	8
Farmer	9	11
Housework	14	6
Non-working time	15	5

ORS: (a) The severest criticism that can be made of a farmer in Homestead is that he neglects his fields during the farming season.

(b) During the farming season the farmers believe they should always work from dawn until dark, and if the work is not finished they believe they should work at night by their tractor lights.

(c) On every day of the year in Homestead (except during the busy farming season) the observer will find groups of men loafing and talking at the repair shops, groups of women loafing and talking in the stores. By way of contrast, one never finds groups of loafers in the stores in the Mormon community.

(d) Homestead men often spend an entire day loafing and drinking beer at the bar.

(e) During the winter off-season there are frequent all-day poker and "moon" games.

(f) All Homestead stores and shops are provided with seats and benches for the loafers. In the Mormon community of Rimrock there are no seats or benches in the stores or shops. The only time the general store in Rimrock had benches for loafers was when a Texan was hired to manage the store!

(g) The writer found it psychologically impossible to while away the long hours in a loafing group and was almost always the first person to leave these groups.

(h) When a person leaves a group of loafers or a visit in a home the most frequent comment is "Don't rush off," or "What's your hurry?"

(i) The Homesteaders typically arrive before the appointed hour for a party, dinner engagement, or other affair, instead of rushing in at the last minute like a busy urbanite.

(j) The Homesteaders always arise early in the morning (typically by sunrise); if they have farming to do, they go to work; if they don't have work to do, they "get up town" early for a day of loafing.

(k) The Homesteaders always say that one of the main reasons they like living in Homestead is that they "have time to loaf." "You only have to work four months a year here; but if you are raisin' cotton in Texas, you have to work 12 months a year."

(l) There is always much discussion about a few people in Homestead who are shiftless and "don't get ahead"; in the same breath, the Homesteaders go on to criticize the few people who work too hard, who stay

APPENDIX

up all hours during the night, even during the winter, to work, and who don't enjoy life.

Variant Responses
INT: Two informants expressed the point of view that the Homesteaders did not work hard enough and should be busy all the time instead of loafing up town.
ORS: Four older people in Homestead are always working and are uncomfortable when they are not occupied. However, they are a matter of special comment by the rest of the community.

V. Major Value-Orientation: GROUP SUPERIORITY AND INFERIORITY

1. Associated Value: "HERE IN HOMESTEAD WE SHOULD TREAT EVERYBODY AS EQUALS," BUT "WE SHOULD NOT FORGET THAT THE 'WHITE PEOPLE' ARE SUPERIOR TO 'MEXICANS AND INDIANS.'"

Dominant Responses
INT: (a) The value placed upon equality *within* the community of Homesteaders was given concrete expression by all 20 informants interviewed.
(b) The value placed upon the superiority of "white people" was given concrete expression by 16 of the 20 informants interviewed.
COQ: The questionnaire did not deal with problems of intergroup relationships.
ORS: (a) Although differences in wealth and family background exist within the community, strong feelings are expressed to the effect that "we are all plain old bean-farmers."
(b) In the larger community affairs (such as dances and "42" parties) every person in the community is invited, and everyone feels that one should participate in these affairs with everyone else, including the residents of the community who appear to be genuine "Tobacco Roaders" to an outside observer.
(c) There is always insistence that *all* the community *should* participate in the selection of committees, delegates, and other decisions involving the total community.
(d) There exists a markedly egalitarian relationship between men and women in the social structure of Homestead.
(e) On the other hand, the Homesteaders maintain feelings of superiority toward the "Mexicans" and "Indians" in the area, these groups being (up to a point) equated with Negroes ("colored people").
(f) The Homesteaders hire the Spanish-Americans and Indians as laborers, but they never work for the Spanish-Americans or Indians.
(g) The Homesteaders and the Spanish-Americans attend each other's dances, but most Homesteaders are opposed to dancing with the Spanish-Americans.

(h) In the early 1930's the Homesteaders raised violent objections to having a Spanish-American teacher in their grade school. The schoolhouse was burned down as a protest, and a separate school was organized in which one of the Homestead men was hired by the families at $5 per month per child to teach school. The Homesteaders also raised objections to sending their teen-age children to the high school in Tapala and eventually managed to have a high school established in Homestead. In 1952 a young, attractive, fair-skinned Spanish-American girl became one of the high school teachers. The students went home after the first day of school and commented to their parents, "Why, she's not black!"

(i) Resentment against the Spanish-Americans still flares up periodically, especially in competitive situations. When the Homestead basketball team was clearly beaten by a superior Spanish-American team in Magdalena in 1950, the Homesteaders attributed their loss to the fact that they "got a raw deal from the Mexican referee," and added that, "you can't expect a fair deal in a town full of Mexicans." At a baseball game against a Spanish-American team in 1952, one of the Texan wives gave a pep talk to the players, saying "Get in there and fight — remember the Alamo!"

(j) In 1947 open conflict with the Spanish-Americans started at a dance when one of the young Texans pushed a Spanish-American at the door of the dance hall and said, "Out of my way, you dirty Mexican." The versions of the ensuing course of events differ. The Texan version is that several "Mexicans" jumped on the young Homesteader and beat him up, and then the Texans had to come to his defense; the Spanish-American version is that the young Texan was beaten in a fair fight, but reported to his fellow Texans that "a whole gang of Mexicans had beaten him up" and that the Texans then proceeded to beat up almost every Spanish-American at the dance. At any rate, the conflict went on for about two weeks in the form of fist-fights which took place whenever the young Homesteader could get his hands on any of the Spanish-Americans who were involved in the original fight.

(k) The Homesteaders generally avoid intimate social contacts with the Spanish-Americans and Indians, such as eating meals with them.

(l) Sexual intercourse and intermarriage with Spanish-Americans and Indians is strongly opposed by most Homesteaders. No intermarriage has taken place between the Homesteaders and Indians, but there have been two cases of intermarriage locally with Spanish-Americans. Even though these marriages were with girls from the leading family of Tapala, there were (and continue to be) strong objections to these marriages, and the children are regarded as "half-breeds."

(m) There is continuing resentment about Spanish-American control of politics in the county and state. When three of the officials at the polls in 1950 were Spanish-Americans, there were loud objections voiced in Homestead. In this same election Homestead voted against a Spanish-American who had actually done a great deal for the community in helping to get roads built, etc., and voted instead for a Texan by the name

APPENDIX

of Gene Autry who was not even known personally in Homestead and had done nothing for the community.

(n) The Texans consider their own cultural patterns to be so superior to those of Spanish-Americans and Indians that they make no effort to become familiar with their customs. Despite over 20 years of residence in the area, no Homesteaders (except for the two married to Spanish-Americans) know anything about the patron saint (an important aspect of the culture of all Spanish-American villages) in Tapala. Only a handful of the Homesteaders have attended the famous Pueblo Shalako; the others have never attended and express little interest in going. Furthermore, Indians and Spanish-Americans are constantly criticized for not trying to learn English while at the same time the Texans make little effort to learn Spanish, Pueblo, or Navaho.

(o) Indians are frequently charged higher prices at Homestead stores than "white people."

(p) In 1949 the Navaho veterans were excluded from the VA farm training class on the excuse that they were "too much trouble."

(q) There have been two Spanish-American burials and one Navaho burial in the Homestead cemetery. These graves are carefully separated from the Homesteaders' graves, with the apparent intention of maintaining the social distance in death as in life.

(r) The Spanish-Americans and Indians who are liked best by the Homesteaders are those who are most highly acculturated, or, in other words, those who have been most willing to grant that the "white peoples'" ways of life are the best and to change their own patterns accordingly. For example, the Homesteaders have the highest praise for "Joe Deerfoot," an Indian who lives in Gallup, speaks good English, and is a good Baptist.

Variant Responses

INT: Four of the informants expressed the point of view that the Spanish-Americans and Indians were just as good as "white people" and were being unjustly treated.

ORS: Eight families in the community (including the two men who are married to Spanish-American women) manifest more tolerant attitudes toward Spanish-Americans and Indians.

2. Associated Value: "HOMESTEAD IS A BACKWOODS TOWN AND WE SHOULD TRY TO MAKE THE COMMUNITY AND OUR WAYS OF LIVING MORE LIKE CITY WAYS."

Dominant Responses

INT: This value was given concrete expression by 15 of the 20 informants interviewed.

ORS: (a) The Homesteaders frequently speak of Homestead as "the jumping-off place."

(b) It is constantly necessary for the field workers to reassure the Homesteaders that they need not apologize for conditions in Homestead.

The Homesteaders apologize for almost everything from their homes to their manners. They are sensitive about their speech habits, food habits, and dress habits when in contact with urbanites.

(c) The Homesteaders manifest longings to do and to have the things that appeal to the typical American middle-class urban families. The women comment on their "rough hands" while playing cards; they speak of "dream houses," clothes, hairdos, etc. They want to learn to play bridge and canasta. The men wonder "what Harvard thinks of our old working clothes."

(d) The Homesteaders expressed the point of view that "it's sure nice of you college people to come out here and study us hicks," but they often expressed feelings of inadequacy about having enough education or being intelligent enough to understand our questions.

(e) The Homesteaders place great emphasis upon their baseball playing as a means of "putting Homestead on the map" and making the larger world realize that there is a growing community at Homestead.

(f) The Homesteaders always remember outsiders from the city, whom they may have met only once, even when the outsiders do not remember the residents of Homestead they met.

(g) When they visit the city, the Homesteaders manifest feelings of inferiority and helplessness. For example, one time a family got lost in Los Angeles and took six hours to find their way out of the city. In general, the travel experience of the Homesteaders is very limited.

(h) Feelings of inferiority extend also to the large ranchers who have more wealth and a more secure position in the outside world, and even, to some extent, to the nearby Mormon community which was settled before Homestead and possesses more wealth and higher status in the area. One of the Homesteaders reported that when the Mormon baseball team came to Homestead, the members of the Homesteader team "used to slip off and not invite them to our homes because we didn't want them to see the kind of houses we live in."

Variant Responses

INT: Five of the informants interviewed expressed the point of view that "our country ways are best for us."

NOTES

NOTES

INTRODUCTION

1. See Clyde Kluckhohn, "The Limitations of Adaptation and Adjustment as Concepts for Understanding Cultural Behavior," in *Adaptation* (ed., John Romano; Ithaca: Cornell University Press, 1949), pp. 99–113.

2. I should like to acknowledge the stimulation I have received from my colleagues in the Department of Social Relations, especially from the work of Clyde and Florence Kluckhohn and of Talcott Parsons. Even though at times I depart quite radically from the conceptual definitions and theoretical emphases of these writers, my thinking in the preparation of this theoretical section has been influenced by their recent publications (see C. Kluckhohn in *Adaptation*, and "Values and Value-Orientations in the Theory of Action," in *Toward a General Theory of Action* [eds., T. Parsons and E. Shils; Cambridge, Mass.: Harvard University Press, 1951], pp. 388–433; Florence R. Kluckhohn, "Dominant and Substitute Profiles of Cultural Orientations: Their Significance for the Analysis of Social Stratification," *Social Forces*, vol. 28, no. 4 (1950), pp. 376–393; Talcott Parsons, *The Social System* [Glencoe, Ill.: The Free Press, 1951]).

3. See John M. W. Whiting and Irvin L. Child, *Child Training and Personality: A Cross-Cultural Study* (New Haven: Yale University Press, 1953), pp. 63–64.

4. See especially Margaret Mead, *Male and Female: A Study of the Sexes in a Changing World* (New York: William Morrow and Co., 1949).

5. See C. Kluckhohn in *Adaptation*.

6. See Parsons, *The Social System*, pp. 26–36; see also D. F. Aberle, A. K. Cohen, A. K. Davis, M. J. Levy, Jr., and F. X. Sutton, "The Functional Prerequisites of a Society," *Ethics*, vol. 60, no. 2 (1950), pp. 100–111.

7. For example, "culture" has been described in terms of different types of patterns — systemic, total-culture, etc. (Alfred L. Kroeber, *Anthropology* [New York: Harcourt, 1948], pp. 311–318); as a cluster of patterns on various levels of abstraction (Ralph Linton, *The Cultural Background of Personality* [New York: Appleton-Century, 1945], pp. 27–54; C. Kluckhohn, "Covert Culture and Administrative Problems," *American Anthropologist*, vol. 45 [1943], pp. 213–227; Alfred L. Kroeber and Clyde Kluckhohn, *Culture: A Critical Review of Concepts and Definitions*, Peabody Museum of Harvard University, Papers, vol. 47, no. 1 [1952], pp. 167–171); as a body of customs (John Gillin, *The Ways of Men: An Introduction to Anthropology* [New York: Appleton, 1948], p. 181); as a set of "conventional understandings" (Robert Redfield, *The Folk Culture of Yucatan* [Chicago: Univer-

sity of Chicago Press, 1941], p. 132); as a roster of traits (see Leslie Spier, *The Sun Dance of the Plains Indians: Its Development and Diffusion*, American Museum of Natural History, Anthropological Papers, vol. 16, no. 7 [1921], pp. 451–527); or as a kind of superorganic reality with laws of its own (Leslie A. White, *The Science of Culture: A Study of Man and Civilization* [New York: Farrar, 1949], pp. 121–145).

8. This directionality in cultural processes looms importantly in accounting for the differences between cultural groups which exist at approximately the same technological level, live in similar physical environments, and are exposed to similar historical influences. It is comparable to the kind of phenomenon that Edward Sapir (*Language: An Introduction to the Study of Speech* [New York: Harcourt, Brace and Co., 1921], p. 60) describes with the concept of *drift* in linguistic materials and that Fred Eggan ("Some Aspects of Culture Change in the Northern Philippines," *American Anthropologist*, vol. 43 [1941], pp. 11–18) has described in cultural materials from the Northern Philippines. See also Kroeber and Kluckhohn, *Culture: A Critical Review*, p. 189, and Melville J. Herskovits, *Man and His Works: The Science of Cultural Anthropology* (New York: Knopf, 1948), pp. 581–588 for lucid discussions of the phenomenon of drift in culture.

9. F. Kluckhohn, "Dominant and Variant Cultural Value Orientations," in *Social Welfare Forum, 1951* (New York: Columbia University Press, 1951).

10. C. Kluckhohn in *Toward a General Theory* (eds., Parsons and Shils), p. 411.

11. I am indebted to Gordon Allport who suggested this highly useful instrument.

12. This profile of value-orientations is based upon a modified version of Florence Kluckhohn's illuminating conceptual scheme for the treatment of key orientations cross-culturally (see F. Kluckhohn in *Social Forces*, vol. 28).

13. See Appendix for a summary of the basic data from which these conclusions about the Homesteaders' value-orientations were drawn.

14. It should be noted that in this study the methodological approach to these problems is different from that in my monograph on *Navaho Veterans: A Study of Changing Values*, Peabody Museum of Harvard University, Papers, vol. 41, no. 1 (1951). In both cases the research concerns the problem of the relationship of values to other determinants in the processes of change. However, in *Navaho Veterans* a series of individuals were studied as focal points for the analysis of change. In the present study, I am concerned with individual behaviors only in so far as these provide data for abstraction and inference about larger social and cultural processes. In other words, individual behaviors are to be regarded as reference rather than focal points in the analysis.

15. See especially Jurgen Ruesch and Gregory Bateson, "Structure and Process in Social Relations," *Psychiatry*, vol. 12, no. 2 [1949], pp. 105–124.

NOTES: INTRODUCTION

16. Ralph Linton, *The Study of Man* (New York: Appleton-Century, 1936), p. 296.

17. It might be noted that a processual approach is in line with recent physical-science thinking on the origin and development of the universe and with biological-science thinking on the origin and development of life forms (see, for example, George Gamow, "The Origin and Evolution of the Universe," *American Scientist*, vol. 39, no. 3 [1951], 393–406, and George G. Simpson, *The Meaning of Evolution* [New Haven: Yale University Press, 1949]; see also Lancelot Law Whyte, *The Next Development in Man* [New York: New American Library, 1950], and Philip H. Bagby, "Culture and the Causes of Culture," *American Anthropologist*, vol. 55, no. 4 [1953], pp. 535–554).

18. Cf. the findings of Julian H. Steward ("Cultural Causality and Law: A Trial Formation of the Development of Early Civilization," *American Anthropologist*, vol. 51 [1949], pp. 1–27) on the comparable initial cultural developments and subsequent changes in the early civilizations of the world. See also Gordon R. Willey, *Prehistoric Settlement Patterns in Viru Valley, Peru*, Bureau of American Ethnology, Bulletin, no. 155 (1953).

19. See Benjamin D. Paul, "Mental Disorder and Self-Regulating Processes in Culture: A Guatemalan Illustration," *Interrelations Between the Social Environment and Psychiatric Disorders* (New York: Milbank Memorial Fund, 1953), for an illuminating use of the concept of culture as a "self-correcting" mechanism in a case of mental disorder.

20. See Ray Allen Billington (with the collaboration of Hames Blaine Hedges), *Westward Expansion: A History of the American Frontier* (New York: Macmillan, 1949).

21. See United States Congress (House of Representatives), *Hearings before the Select Committee to Investigate the Interstate Migration of Destitute Persons* (John H. Tolan, Chm.), 76th Congress, 3rd Session (1940); Sigurd Johansen, "Migratory-casual Workers in New Mexico," in *Migratory Cotton Pickers in Arizona* (by Malcolm Brown and Orin Cassmore, Works Projects Administration, Division of Research, 1939), Appendix A, pp. 83–91; C. D. Lively and Conrad Taeuber, *Rural Migration in the United States*, Research Monograph, no. 19 (1939), Works Projects Administration, Division of Research; Rupert Vance, *All These People* (Chapel Hill: University of North Carolina Press, 1945).

22. Seymour J. Janow, "Volume and Characteristics of Recent Migration to the Far West," *Hearings before the Select Committee to Investigate the Interstate Migration of Destitute Persons*, part 6, 76th Congress, 3rd Session (1940), pp. 2270–2281.

23. United States Congress, *Tolan Committee Hearings*, part 5, pp. 1762–1764; see Wilfrid C. Bailey, "A Study of a Texas Panhandle Community: A Preliminary Report on Cotton Center, Texas" (unpublished MS, at Values Study Office, Laboratory of Social Relations, Harvard University, 1951), for a description of a community on the South Plains in the Texas Panhandle.

24. Janow, *Interstate Migration Committee Hearings*, pp. 2270–2281.
25. See Walter Goldschmidt, *As You Sow* (New York: Harcourt, 1947), and Carey McWilliams, *Factories in the Field: The Story of Migratory Farm Labor in California* (Boston: Little, Brown and Co., 1939).
26. Orval E. Goodsell, *An Economic Appraisal of Dry-Land Farming on the Zuni Plateau, New Mexico* (United States Department of Agriculture, Bureau of Agricultural Economics, 1943), p. 11.
27. Cf. D. Harper Simms, "Dust Bowlers Get a Third Chance," *Land Policy Review*, vol. 4, no. 12 (1941), pp. 11–14.
28. See Chapter 6 for additional details on this multicultural setting.
29. See pp. 55–56 for a description of these market towns.

CHAPTER 1

THE GEOGRAPHY AND DEMOGRAPHY OF HOMESTEAD

1. The landscape of the nearby Rimrock area has been described in John L. Landgraf, *Land Use in the Ramah Area of New Mexico: An Anthropological Approach to Areal Study*, Peabody Museum of Harvard University, Papers, vol. 42, no. 1 (1954).
2. I am indebted to Mr. Tom O. Meeks, Soil Conservation Service, Albuquerque, New Mexico, for providing the geological data for this section.
3. For further information on these plants, see Paul A. Vestal, *Ethnobotany of the Ramah Navaho*, Peabody Museum of Harvard University, Papers, vol. 40, no. 4 (1952), who lists the flora of the nearby Rimrock Navaho area.
4. For further identification of these animals, see Vernon Bailey, "Mammals of New Mexico," *North American Fauna*, no. 53 (United States Department of Agriculture, Bureau of Biological Survey, 1931); for bird identifications, see Florence M. Bailey, *Birds of New Mexico* (Santa Fe: New Mexico Department of Game and Fish, 1928).
5. See especially J. H. Dorroh, Jr., *Certain Hydrologic and Climatic Characteristics of the Southwest*, University of New Mexico Publications in Engineering, no. 1 (Albuquerque: The University of New Mexico Press, 1946), from which the following facts on the general climatic patterns in the Southwest were drawn.
6. Dorroh, pp. 1–4.
7. Homestead's rainfall may have been slightly higher than that of Q in the years reported, but the figures from EMA and EMM are undoubtedly close to the precipitation in Homestead for the years reported.

Inasmuch as Homestead's beans were sold through a number of different business establishments in two or three market towns before the local bean warehouse was opened for business in 1940, no accurate figures on total bean production are available for the period 1932 to 1940. The figures on average yield per acre for this period are from Goodsell, *An*

NOTES: CHAPTER 2

Economic Appraisal, p. 37. A particularly dry summer accounts for the low yield in 1939. In other years production was also influenced by the distribution of the total rainfall during the year, but accurate figures are not available on this distribution for the immediate area of Homestead. Hence, only totals are given here as an approximation of moisture conditions.

The figures on total bean production for the years 1941–1950 are taken from the records of the local bean warehouse in Homestead. The actual totals from the community would be somewhat higher inasmuch as a few farmers sell their beans through other warehouses located in the market towns. However, the relative amount of bean production from year to year can be judged accurately from this table. Since accurate figures are not available for cultivated land devoted to beans from 1941 to 1948, no estimate of average yield per acre can be ventured for this period. It is known, however, that the yield per acre in 1949 was approximately 2.5 sacks. In 1947 the yield was markedly reduced by a killing frost on September 8.

8. See Landgraf, *Land Use in the Ramah Area*, for a discussion of the problem of land-use in neighboring areas.

9. Occasionally there may be crop damage from severe hailstorms. For example, in the summer of 1952, one fourth of Homestead's farms suffered 30 to 100 per cent damage from a hailstorm on July 27th.

10. See Vestal, *Ethnobotany of the Ramah Navaho*.

11. Although the difference between immigration and emigration leaves a balance of 32 families, it must be remembered that other families originate in Homestead from marriages occurring within the community. Hence, there was a total of 61 families in Homestead in 1949.

Chapter 2

THE ECONOMY OF HOMESTEAD

1. The holders of these grazing permits pay $.12 per cow per month for the use of the land. In the Homestead area, operators are permitted to graze eight or ten cows per section; thus the annual cost of using a section runs from $11.52 to $14.40 — a very small amount for the right to use a section.

2. Only one man in the community works as a full-time hired hand.

3. These absentee-owners are persons who once lived in Homestead and have since moved away.

4. Barns are also lacking in Cotton Center, the ancestral area of the Homesteaders (Cf. Bailey, "Study of a Texas Panhandle Community").

5. There was some trapping of coyotes, foxes, and wild cats in the 1930's. Another activity which might be defined as collecting is the digging up of prehistoric pueblo pottery which was sold to museums back in Texas in the early 1930's.

6. In addition to letters from relatives and friends, bills from the city, and packages from mail-order houses, the Homesteaders also receive the following newspapers and magazines via the United States mail:

Newspapers	
Daily newspapers from New Mexican cities	5
Weekly newspapers from New Mexican cities	44
Daily newspapers from Texan cities	3
Farm Journals	
Western Farm Life	42
The Farm Journal	40
The Country Gentleman	30
New Mexico Agriculture	20
New Mexico Farmer Stockman	2
Capper's Farmer	1
Other Magazines	
Reader's Digest	4
Colliers	4
Woman's Home Companion	3
True Story	3
Life	3
Time	2
American Magazine	2
Saturday Evening Post	1
Look	1
Better Homes and Gardens	1
Wee Wisdom	1

The combination store and *café* sells 22 copies of *Grit: America's Family Newspaper* each week, and when the drug store was in operation its news-stand sold the following newspapers and magazines:

Newspapers	
Los Angeles *Examiner* (Sunday edition)	6
Los Angeles *News* (Sunday edition)	4
Magazines	
Western Stories	8
Woman's Home Companion	6
Colliers	6
Look	6
American Magazine	3

7. See Irving Telling, "New Mexican Frontiers: A Social History of the Gallup Area, 1881–1901" (unpublished Ph.D. dissertation, Harvard University, 1952), for additional details on Gallup.
8. Goodsell, *An Economic Appraisal*, p. 16.
9. Goodsell, p. 50.

10. Goodsell, p. 6.
11. There are significant differences in average yields between the fields of the more experienced farmers and the fields of less experienced operators.
12. The net income was derived by deducting farm expenses (cost of fuel, seed, taxes, and depreciation and repairs on machinery) and business expenses (rent, fuel, taxes, extra labor, and electricity), excluding the value of unpaid family labor, from gross farm and business income. In other words, net income is the compensation, including the value of family living expenses, from the farm or service business in return for capital, labor, and management of the operator, and labor of unpaid members of the operator's family.

The average costs per acre of producing beans in 1949 were calculated as follows.

Item	Per acre cost
Fuel and oil	$1.20
Seed, 15 pounds at 7¢ a pound	1.05
Depreciation on machinery	2.00
Repairs on machinery	.35
Taxes on land	.03
TOTAL	$4.63

The "town" families include the eleven families who live in Homestead and derive their incomes predominantly from goods and services provided for the community and two persons who live on old-age pensions and/or relief payments. There are also fourteen farm families who derive partial incomes from part-time service activities, but these incomes are added to their farming incomes in this table.

CHAPTER 3

HOPEFUL MASTERY OVER NATURE

1. Goodsell, *An Economic Appraisal*, p. 14.
2. In 1936, the Farm Security Administration also loaned money to the Homestead Water Coöperative Association to drill a community well.
3. Goodsell, p. 14.
4. Many local ranchers and businessmen contend that the "New Deal" government assistance kept the community going in the 1930's, and that without it Homestead never would have survived the depression. These same men, who also express the view that "New Deal" funds were needlessly and uneconomically expended, received very substantial government assistance during the depression.
5. See Evon Z. Vogt and Thomas F. O'Dea, "A Comparative Study of the Role of Values in Social Action in Two Southwestern Communities," *American Sociological Review*, vol. 18, no. 6 (1953), pp. 645–654.

6. The board is composed of representatives from the Bureau of Agricultural Economics, the Farm Security Administration, the Soil Conservation Service, the Agricultural Adjustment Administration, the Grazing Service, the General Land Office, the Indian Service, and the Office of Land Utilization.
7. Goodsell, p. 3.
8. Goodsell, p. 33.
9. Indeed, the Homesteaders had not seen Goodsell's final report until the writer brought a copy to the community in 1949.
10. The knife-sled was developed by one of the local farmers on the basis of his previous experience in other dry-farming areas. It does a more effective job of cutting the weeds and turning the soil up around the young bean plants than some of the expensive, commercially-made machines.
11. The most expensive item of machinery is the farm tractor, and in 1950 there were fifty-one in Homestead. Twenty-one were John Deere models ranging in cost from $2300 to $2700. Other types were: Case (eleven), International Farm-All (eleven), and Ford (eight), which was the least expensive and cost $1600. The larger and more popular tractors handle four-row equipment; the smaller types handle only two-row equipment. Also, there were fourteen combines — the next most expensive item, costing from $1800 to $2200; farmers without combines usually had their threshing done by those who owned this implement.

Other items of machinery which represent investments of $500 or more include the combination lister-planters, the cultivators, and the larger discs. The least expensive item is the knife-sled, which is manufactured in the local blacksmith shops at a cost of approximately $50. The total minimum investment in Homestead for a mechanized farmer (without a combine) is approximately $4000, and a few of the owners of larger farms have up to $8000 invested in farm machinery.
12. The annual agricultural cycle is discussed in detail in Chapter 5.
13. Usually three, rather than four, rows are listed at a time, in order to plow out the tracks of the three tractor wheels, with the lister plows placed directly behind each of the wheels. If this is not done, the tractor wheels pack down the soil and make it impossible to use the planter efficiently. The planting is then done four rows at a time in order to plant in the soft furrows left by the listing process.
14. See Goodsell.
15. At present the total number of motor vehicles in the community is 72: 32 pickup trucks, 35 passenger cars, and 5 larger trucks.
16. See especially Bronislaw Malinowski, "Magic, Science, and Religion," in *Science, Religion, and Reality* (ed., Joseph Needham; New York: Macmillan, 1925), pp. 19–84.
17. James West, *Plainville, U.S.A.* (New York: Columbia University Press, 1945), p. 11.
18. C. C. Taylor, *Rural Sociology* (New York: Harper, 1933), p. 145.
19. I am indebted to Mr. James Finkelstein for permission to utilize

NOTES: CHAPTER 3

materials from his Honors Thesis, which was prepared under my direction and deals with the problems of folk ritual in Homestead (see Finkelstein, "A Functional Analysis of the Folk-Ritual System in a Small Agricultural Community" [unpublished Honors Thesis, Harvard College, 1951]).

20. See Evon Z. Vogt, "Water Witching: An Interpretation of a Ritual Pattern in a Rural American Community," *The Scientific Monthly*, vol. 75, no. 3 (1952), pp. 175–186, for a brief summary of the development of this theory of ritual.

21. See Talcott Parsons, "The Theoretical Development of the Sociology of Religion: A Chapter in the History of Modern Social Science," *Journal of the History of Ideas*, vol. 5, no. 2 (1944), pp. 176–190.

22. Alfred L. Kroeber, *Anthropology*, p. 604.

23. A. R. Radcliffe-Brown, *Taboo* (Cambridge, England: Cambridge University Press, 1939), pp. 5–46.

24. George C. Homans, "Anxiety and Ritual: The Theories of Malinowski and Radcliffe-Brown," *American Anthropologist*, vol. 43, no. 2 (1941), pp. 164–172.

25. C. Kluckhohn, *Navaho Witchcraft*, Peabody Museum of Harvard University, Papers, vol. 22, no. 2 (1944), pp. 45–72.

26. For a more complete discussion of water witching, see Vogt in *The Scientific Monthly*, vol. 75. The materials summarized here are drawn from this paper and are reprinted with the permission of *The Scientific Monthly*. See also Thomas M. Riddick, "Dowsing — an Unorthodox Method of Locating Underground Water Supplies or an Interesting Facet of the Human Mind," *Proceedings of the American Philosophical Society*, vol. 96, no. 5 (1952), pp. 526–534, for additional materials on the ritual nature of dowsing.

27. In this respect, dowsing comprises a skill which is acquired by "divine stroke" as in the "shamanistic" tradition, rather than a body of knowledge which is transmitted by training in a "priestly" tradition.

28. Six wells were dowsed by a second witch who lived in the community for a few years, making a total of 24 wells that were located by dowsing.

29. Finkelstein, "Functional Analysis," p. 65.

30. For a description of this study, which was made by Professor Wilfrid C. Bailey of the University of Texas, see Bailey, "Cotton Center, Texas, and the Late Agricultural Settlement of the Texas Panhandle and New Mexico," *The Texas Journal of Science*, vol. 4, no. 4 (1952), pp. 482–486; also, Bailey, "Study of a Texas Panhandle Community."

31. See W. N. White, W. L. Broadhurst, and J. W. Lang, *Ground Water in the High Plains in Texas* (Austin: Texas State Board of Water Engineers, 1940).

32. See Vogt in *The Scientific Monthly*, vol. 75.

33. Mr. Tom O. Meeks has called my attention to the fact that the Soil Conservation Service maintains only two geologists in the Southwest who give advice on locating wells; hence their visits to any given community are necessarily infrequent.

34. For the "signs" the Homesteaders depend mainly upon Dr. J. H. McLean's Almanac (now in its 98th year of publication) which is distributed by the Dr. J. H. McLean Medicine Company of St. Louis.
35. The same sign is used by many of the women to judge the proper time to wean infants.
36. Parsons, *Social System*, pp. 466–469.
37. At the request of the clergy, the governor of Texas set aside a day of prayer for rain in 1952.
38. It is significant that the Navahos and Pueblos also hold this belief.
39. Bailey points out that faith healing is still extremely common in western Texas (see Bailey, "Study of a Texas Panhandle Community").
40. For additional data on this faith healer see K. R. Ellis, "They Call Her Miracle Woman," *True: The Man's Magazine*, vol. 12 (1943), pp. 18–23, 110–112.
41. Ellis, p. 112.
42. Ellis.
43. See Munro S. Edmonson, "Los Manitos: Patterns of Humor in Relation to Cultural Values" (unpublished Ph.D. dissertation, Harvard University, 1952), for a treatment of the relation of humor to tension-producing situations.
44. This behavioral phenomena may be called a type of "proto-magic."
45. Clifford J. Geertz, "Drought, Death, and Alcohol in Five Southwestern Cultures" (unpublished MS, at Values Study Office, Laboratory of Social Relations, Harvard University, 1951), p. 18.
46. Earl H. Bell, *Culture of a Contemporary Rural Community: Sublette, Kansas*, Rural Life Studies, no. 2 (Department of Agriculture, Bureau of Agricultural Economics, 1942), p. 43. See also Jean Burnet's *Next-Year Country* (Toronto: University of Toronto Press, 1951), pp. 4–5, which describes a similar type of gambling orientation toward farming in the dry plains region of Alberta in Canada.
47. Robin M. Williams, Jr., *American Society: A Sociological Interpretation* (New York: Knopf, 1951).

Chapter 4

LIVING IN THE FUTURE

1. Williams, *American Society*, p. 404.
2. Another bean-farming community, Dove Creek, Colorado, is also self-named "Pinto Bean Capital of the World."
3. *Homestead Promoter*, March 3, 1938.
4. See Table III.
5. It is significant that in the same year an identical proposal for the construction of a high school gymnasium was presented to the Mormon community of Rimrock. The plan was accepted by the Mormon group, and the new gymnasium (and high-school building) is now nearing com-

NOTES: CHAPTER 6

pletion, using volunteer labor by the men of the community (*see* Vogt and O'Dea in *American Sociological Review*, vol. 18).
   6. Local chapter of the New Mexico Livestock and Farm Bureau.
   7. See especially W. Lloyd Warner and Paul S. Lunt, *The Social Life of a Modern Community* (New Haven: Yale University Press, 1941), p. 51.
   8. I have added one case of a young woman who had completed high school just prior to her family's arrival in Homestead in the early 1930's.

CHAPTER 5

WORKING AND LOAFING

   1. See Charles A. and Mary R. Beard, *The American Spirit* (New York: Macmillan, 1937), pp. 661–670.
   2. See Jurgen Ruesch and Gregory Bateson, *Communication: The Social Matrix of Psychiatry* (New York: W. W. Norton, 1951). Talcott Parsons has suggested (personal communication, February 1953) that an important shift is taking place in this generalized American value-orientation, with more attention being devoted to "fun" and "recreation" after working hours.
   3. F. Kluckhohn in *Social Forces*, vol. 28, pp. 378–379.
   4. The period from noon until sundown is called "evening"; after dark it is "night."
   5. A scratcher is made of boards through which nails have been driven and is dragged behind the tractor.
   6. Williams, *American Society*, p. 394.
   7. These loafing groups have important integrative functions in the social system, a point that is further discussed in Chapter 7.
   8. F. Kluckhohn in *Social Forces*, vol. 28, pp. 378–379.
   9. See Thomas F. O'Dea, "Mormon Values: The Significance of a Religious Outlook for Social Action" (unpublished Ph.D. dissertation, Harvard University, 1953).
   10. See Evon Z. Vogt, "Town and Country: The Structure of Rural Life," in *Democracy in Jonesville: A Study in Quality and Inequality* (W. Lloyd Warner *et al*; New York: Harper, 1949), pp. 236–265.
   11. Bailey, "Study of a Texas Panhandle Community," p. 39.
   12. Bell, *Culture of a Contemporary Rural Community*, no. 2, p. 53.
   13. West, *Plainville, U.S.A.*, pp. 99–107.

CHAPTER 6

GROUP SUPERIORITY AND INFERIORITY

   1. See John Otis Brew and Edward Bridge Danson, Jr., "The 1947 Reconnaissance and the Proposed Upper Gila Expedition of the Peabody Museum of Harvard University," *El Palacio*, vol. 55, no. 7 (1948), pp.

211–222; Watson Smith, "Preliminary Report of the Peabody Museum Upper Gila Expedition, Pueblo Division, 1949," *El Palacio*, vol. 47, no. 12 (1950), pp. 392–399; E. B. Danson, "Preliminary Report of the Peabody Museum Upper Gila Expedition, Reconnaissance Division, 1949," *El Palacio*, vol. 57, no. 12 (1950), pp. 383–391; and Charles R. McGimsey III, "Peabody Museum Upper Gila Expedition: Pueblo Division Preliminary Report: 1950 Season," *El Palacio*, vol. 58, no. 10 (1951), pp. 299–312.

2. For additional details on this cultural setting and the history of the settlement, see Edmonson, "Los Manitos"; O'Dea, "Mormon Values"; John M. Roberts, *Three Navaho Households: A Comparative Study in Small Group Culture*, Peabody Museum of Harvard University, Papers, vol. 40, no. 3 (1950); Telling, "New Mexican Frontiers"; and Vogt, *Navaho Veterans*, pp. 11–17.

3. See Chapter 7.

4. See Bailey, "Study of a Texas Panhandle Community."

5. F. Kluckhohn, "Los Atarqueños: A Study of Patterns and Configurations in a New Mexican Village" (unpublished Ph.D. dissertation, Radcliffe College, 1941), pp. 242–248.

6. More detailed data on the interrelationships between the Homesteaders and the Spanish-Americans and other cultural groups will be presented in the writer's forthcoming chapter on "Inter-Cultural Relations" in the final report on the Values Study project.

7. See Vogt, *Navaho Veterans*.

8. Landgraf, *Land Use in the Ramah Area*.

9. Cf. Geertz, "Drought, Death, and Alcohol" for a discussion of these contrasting relationships to the outside world.

10. "Nester" is a term with derogatory connotations, used by ranchers in describing a homesteader.

11. Goodsell, *An Economic Appraisal*, pp. 24ff.

Chapter 7

THE ATOMISTIC SOCIAL ORDER

1. This stress upon individual independence is also clearly documented in Wayne W. Untereiner, "Self and Society: Orientations in Two Cultural Value Systems" (unpublished Ph.D. dissertation, Harvard University, 1952), pp. 92–109. He characterizes Homestead as a "self-oriented" culture and draws an important contrast with the "society-oriented" Pueblo culture of the Southwest.

2. Talcott Parsons and Edward A. Shils, *Toward a General Theory of Action* (Cambridge, Mass.: Harvard University Press, 1951), pp. 190–191.

3. See Lowry Nelson, *The Mormon Village: A Pattern and Technique of Land Settlement* (Salt Lake City: University of Utah Press, 1952), for

NOTES: CHAPTER 7

a discussion of the "isolated farmstead" and "compact village" patterns of settlement.

4. The socialization process will be described in detail in a forthcoming manuscript by Barbara Chartier Ayres on "Child Training in a Texan Village with Special Emphasis on Internalization of Values." Here we are concerned only with the major outlines and a few of the relevant details of the process.

5. See Fred L. Strodtbeck, "Husband-Wife Interaction over Revealed Differences," *American Sociological Review*, vol. 16, no. 4 (1951), pp. 468–473, for experimental evidence of this egalitarian relationship.

6. Cf. Geertz, "Drought, Death, and Alcohol," p. 83.

7. The terms "first cousin," "second cousin," etc., are used, but the Homesteaders are unfamiliar with the terms "first cousin, once removed," etc.

8. F. Kluckhohn, "Los Atarqueños," p. 239.

9. See especially O'Dea, "Mormon Values."

10. Bailey ("Study of a Texas Panhandle Community," p. 41) states: "Religion plays a very important part in the daily lives of the people. Cotton Center may be included in the 'Bible Belt' that covers most of the South."

11. One of these Texan husbands has become a Catholic; the other has not.

12. Alternatively, these "factions" might be called "cliques," but the writer prefers the former term because (a) factions spread across all the age-grades in the community, whereas "cliques" imply sets of persons in the same age-grade; and (b) the word "faction" implies conflict with others, which definitely exists in Homestead.

13. Over a period of time the composition of factions is, of course, also affected by territorial mobility, but this factor does not account for factional realignments which occur while the population is stable.

14. This relationship between factions and churches has also been observed by Wilfrid C. Bailey in his study of Cotton Center, Texas ("The Status System of a Texas Panhandle Community," *The Texas Journal of Science*, vol. 5, no. 3 [1953], pp. 326–331).

15. This man may be described as a "grass roots communist." He has had no connection with or indoctrination from the Communist Party; rather, he has been led to his convictions by feelings of deprivation as he compares his meager land holdings with those of the "big ranchers" who surround his homestead, and by a general newspaper and magazine knowledge of the aims of communism.

16. See Warner and Lunt, *Social Life of a Modern Community*.

17. Cf. West's analysis of Plainville in which the class system is described as a "super-organization" (*Plainville, U.S.A.*, p. 115). It is possible that a community below a certain size in America is characterized by a factional system which forms the nucleus for and develops into a class system as the community grows larger in size.

18. Only six persons whom I met during the course of the field work in Homestead held these attitudes.
19. Cf. F. Kluckhohn in *Social Welfare Forum, 1951.*

Chapter 8

CONCLUSIONS

1. See Vogt, *Navaho Veterans,* pp. 113–114.
2. Alexander A. Goldenweiser, "Loose Ends of Theory on the Individual, Pattern, and Involution in Primitive Society," in *Essays in Anthropology in Honor of A. L. Kroeber* (ed., Robert H. Lowie; Berkeley: University of California Press, 1936), pp. 102–103.
3. Kroeber, *Configurations of Culture Growth* (Berkeley: University of California Press, 1944), p. 763.
4. Walter B. Cannon, *The Wisdom of the Body* (Philadelphia: W. W. Norton Co., 1939), p. 24.
5. See especially Norbert Wiener, "Cybernetics," *Scientific American,* vol. 179, no. 5 (1948), pp. 14–19, and *The Human Use of Human Beings* (New York: Houghton Mifflin Co., 1950).
6. Dickinson W. Richards, "Homeostasis Versus Hyperexis: Or Saint George and the Dragon," *The Scientific Monthly,* vol. 77, no. 6 (1953), p. 294.
7. This fact is particularly evident in the Homesteaders' failure to perceive the long-range decline of the community. For example, in 1950 when my census listed a total population of 232, many of the best-informed residents of Homestead were convinced that the population was at least 350.

Appendix

1. This classification of the key aspects of the life situation is a modified version of Florence Kluckhohn's formulation for the classification and description of important cultural orientations (see F. Kluckhohn in *Social Forces,* vol. 28). These key aspects of human life are not to be regarded as the only important areas of value-orientations. Rather, they are to be regarded as the *most relevant* aspects for the purpose of the central problems under investigation in this study.
2. A copy of the interview schedule is on file in the Values Study office at Harvard University.
3. This questionnaire was designed by Florence Kluckhohn, Fred Strodtbeck, and Kimball Romney and administered by Florence Kluckhohn, John M. Roberts, Fred Strodtbeck, and Kimball Romney with some assistance by the writer. I am indebted to these other project workers for permission to utilize the Homestead results of this questionnaire, a copy of which is on file in the Values Study office and will be published

in F. Kluckhohn, Fred Strodtbeck, John M. Roberts, and Kimball Romney, *Variations in Value-Orientations: A Theory Tested in Five Cultures* (forthcoming).

4. See F. Kluckhohn in *Social Forces*, vol. 28, and in *Social Welfare Forum, 1951*, for the basic discussion of these concepts of dominant and variant value-orientations.

INDEX

INDEX

Adolescence, 146
Adultery, 150, 157
Age, categories, terms used, 145–146; distribution, 35–36; roles, 146–147
Agricultural Adjustment Administration, 65
Agriculture, U. S. Department of, 46, 66, 68–70. *See also* names of agencies
Albuquerque, New Mexico, 19, 56
Almanacs, use of, 86
Anxiety, about drought, 89–90; related to ritual, 75–77, 84, 86; related to water witching, 83–85
Atheists, 161, 167, 172
Authority, adult, 146; rejection of "the law," 140, 155, 157–159

"Bachelor" defined, 147
Bailey, Wilfred C., 119
Baptists, 87–88, 161–164, 167, 170; church controlled by kin group, 152; organized as protest, 166; support of preacher, 54
Baseball team, 96, 116–117
Bell, Earl H., 91, 119
Bible, books of Moses, 78; Isaiah, 88; Matthew, 88
Birth control, 75
Boasting, 91
Boosters Club, 95–96
Buildings, 47–48

California, migrants in, 16
Campbellites, 161
Cannon, Walter B., 184–185
Catholicism, 130
Catholics, 161–162, 164
Childbirth, 75
Civil Works Administration, 65
Climate, 27–30; and income, 57–58; problems for farming, 110–112
Clothing, 49–51
Collecting, pinyon nut, 51
Committees, 97–98
Communication, "barbed-wire" telephone, 71; importance of postmistress, 144
"Communist, grass roots," 167
Community Church, *see* Presbyterians
Community well, 47, 54–55
Conflict situations, use in values research, 10–11
Converts, religious, 138, 164
Coöperation, extent of, 140; in farming, 70, 72; in kin groups, 152–153; within factions, 165
Coöperative Association, Homestead, 54–55, 170
Coöperatives, opposition to, 9, 67, 97
Cotton Center, Texas, 81–82, 119
Credit, 46
"Crime" defined, 158–159
Crises, handling of, 159; stimulate coöperation, 140; use in values research, 9–10
Crops, 43–44; marketing of, 45–46

Dancing, 117; and "equality," 124; and factionalism, 170; occasions for fights, 128; Mormon, 138; Baptist sermon against, 163–164
Debts, 71. *See also* Credit; Loans
Deputy sheriff, 155–157, 159
Disease, inoculations against, 75
Divorce, 157
Domesticated animals, 44, 71
Drinking, 115, 148; the bar, 53, 115–116; "salty dog," 117; and sex division, 149–150
Drought, impetus for migration, 16, 18; anxiety about, 89–90

Economic resources of first settlers, 17
Education, aspirations, 104–106; attitudes toward, 154–155
Emigration from Homestead, 33–34, 60, 101–102; and occupation, 102–104
"Enabling Act," 38
Eskimos, 76
Exchange labor, 47, 70, 152

Factions, 164–171
Faith healers, 10, 88–89
"Failures" defined, 100–101; among young people, 104
Family, extended, *see* Kin groups; Kinship

230 INDEX

Family, nuclear, as basic social unit, 150; role in social system, 142; size, 35
Farm Bureau, 97–98, 170
Farm Security Administration, 65–66
Fauna, 25–27; uses of, 32, 51
Fighting, 128–129; code, 158
Fishing, 51
Flora, 24–25; uses of, 31, 44; gathering of pinyon nuts, 51; witching sticks, 78
Food, 48–49
Friends, credit extended to, 144

Gallup, New Mexico, 19, 55; bankers, 67–68
Gambling, attitude toward farming, 63, 90–91, 99
Game warden, 155
Goldenweiser, Alexander A., 184
Goodsell, Orval E., 17–18, 57, 58, 65, 137
Gossip, importance in values research, 11; as a social control, 158; men's, 115, 158; women's, 158
Government aid, 47, 59, 65–67, 97; opposition to, 66–67. *See also* Loans
Government price supports, 45
Grants, New Mexico, 19, 56

Holiness Church, 161
Homans, George C., 76–77
Homestead Act of 1916, *see* Stock Raising Homestead Act
Homestead Promoter, 95–96
Hunting, 51, 112; code, 159; laws, view of, 155

Illness, 74–75; treatment of, 75. *See also* Faith healers
Income, 51–60; discussed freely, 100
Indian ceremonials, attendance at, 133
Indians, attitudes toward, 125, 130–131. *See also* Navahos; Pueblos
Interior, U. S. Department of the, 39, 41, 68, 136. *See also* names of agencies

Jessel, Mrs. Susie, *see* Faith healers
Jonesville, 119
Justice of the Peace, 155–157

Kin groups, as basis of factions, 166; exchange labor in, 47, 70
Kinship, behavior, 144, 150–153; terms, 151, 153
Kluckhohn, Clyde, 7, 76–77
Kluckhohn, Florence, 7, 110, 156
Kroeber, Alfred L., 76, 184

Labor, division of, 110, 147
Land, value of, 57; methods for acquiring, 37–40; settlement pattern, 6; total area, 40–41; acreage under cultivation, 44; size of farms, 42; self-ownership, 43; laws, compliance with, 157; conflicts over, 136–137
Landgraf, John, 134
Leaders, "real" and "meddlers," 159–160; role in intercultural relations, 128
Leadership, formal positions, 169
Loafing, 109–116, 118–121
Loans, bank, 46, 67–68, 99; government, 45–46
Los Lunas, New Mexico, 56, 156

Magical practices, *see* Ritual
Malinowski, Bronislaw, 75, 76
Marriage, 146–147; cross-cultural, 129–130, 133; and "equality," 124; in-law relationship, 153; laws, compliance with, 157
Mechanization, 68–74
Mental illness, 10, 159
Methodists, 161–162
Mexicans, *see* Spanish-Americans
Migration, to Homestead, 15–18, 32–33; westward, 15–16. *See also* Emigration from Homestead
Missionaries, Mormon, 123, 138; Present-Day Disciple, 164; Seventh-Day Adventist, 164
Moon, phases of, as guide, 81, 85–86
Mormons, local cultural group, 2, 123; orientation to work, 113; activism, 118–119; security derived from church hierarchy, 134–135; relationships with Homesteaders, 137–139; settlement pattern, 142; living in Homestead, 161, 164; preaching in Community Church, 162
Music, 117, 181

INDEX

Navahos, local cultural group, 2, 123; ceremonial development, 5; employed, 51, 70; and cars, 73; witchcraft, 77; orientation to time, 94; on road improvement, 98; in Mormon store, 118; relationships with Homesteaders, 130–132
Nazarenes, 161
Negroes, 122, 125, 129
Neighborhoods, 153–154
Neighbors, credit extended to, 144; exchange labor with, 70
"Nesters," 136
"New Deal," 64–68
New Mexico Department of Public Welfare, 65

Occupations, 44, 51–55; roles, 141–145. *See also* Wagework
O'Dea, Thomas F., 119
Old age, 36, 147
Optimism, 99

Pareto, Vilfredo, 76
Parsons, Talcott, 4, 76
Poker, 113, 118, 148
Political power, ranchers', 137
Politics, 160–161; and "equality," 124
Population, 34–36
Prayer meetings, 87–88
Presbyterians, 161–163, 166–167, 170; support of missionary, 54
"Present-Day Disciples," 161, 164
Prestige, *see* Success, concepts of
Production and Marketing Administration, 45–48, 59, 69
Progress, concept of, 94
Property, self-ownership of, 72–73; and "progress," 94–95; laws, compliance with, 157
Pueblos, local cultural group, 2, 122–123; orientation to nature, 63; and cars, 73; guided by myths, 93; relationships with Homesteaders, 132–134; settlement pattern, 142

Quemado, New Mexico, 19, 56

Radcliffe-Brown, A. R., 76
Rain, word symbol, 89; "rain making," 74
Recreation, activities defined as, 51;

bridge club, 135–136; dominoes, 115, 117–118; Roping Club, 149. *See also* Baseball team; Dancing; Poker; Rodeos
Religion, 161–164; and factionalism, 166; Mormon, attitudes toward, 138; and political power, 169–170; "preachers'" role, 144–145; related to social rank, 167. *See also* Faith healers; Prayer meetings
"Relinquishments," 38
Revival Meetings, 87, 163–164
Richards, Dickinson W., 186
Ritual, concept employed, 77; practices, 63–64, 75–90
Rodeos, 112; Navaho participation in, 131
Role structure, 141

St. Johns, Arizona, 19, 56
Sante Fe Railroad, 38
School, functions of, 155; and factions, 169
Schoolteachers, role, 5–6, 144–145; income, 54; prestige of, 104; Spanish-American, opposition to, 126–127
Seventh-Day Adventists, 161–162, 164
Sex, distribution of, 35–36; relations, 146, 150; roles, 146–150
Social stratification, 123–125, 167–168
Socialization of children, 146–147
Soil, 23–24; erosion, control of, 69
Soil Conservation Service, 65, 70, 83
Spanish-Americans, local cultural group, 2, 123; land holdings, 40; political control by, 56; orientation to nature, 63; employed, 70; electricity in homes, 71; belief in chance, 90; orientation to time, 94; on road improvement, 98; work habits, 113; relationships with Homesteaders, 125–130; place in hierarchy of local groups, 134; settlement pattern, 142; Justice of the Peace, 156
"Squatters," 39–40, 99
Steinbeck, John, 15
Stock Raising Homestead Act, 37; provisions of, 38
Success, concepts of, 100–101, 104
Superstition, *see* Ritual

Taboos, adultery, 157; women in bar, 115
Taylor, C. C., 76
Taylor Grazing Act, 32, 37, 39–41, 66, 136
Tenant farmers, 40, 42–43
Texas, boasting pattern, 91
"Tobacco Roaders," 124, 168
Traditions, "The Loyal Order of Homesteaders," 64
Transportation, 71, 73, 143
Travel, extent of, 135
Trobriand Islanders, 76

Value-orientations, defined, 7
Values, defined, 7
Values Study Project, 2, 9
Veterans, loans to, 46
Veterans Administration farm training program, 54, 102–103; income from, 59; opposition to Spanish-Americans in, 129; opposition to Navahos in, 131

Wagework, outside community, 46; income from, 59–60; for government agencies, 65
Warner, W. Lloyd, 100, 167
Water supplies, 24, 31, 47–48; irrigation attempts, 74; related to witching practices, 80–84; underground, dowser's beliefs about, 79; related to location of farm houses, 85. *See also* Rain: "rain making"
Water witching, 78–85
Weather station, 29, 68
Weber, Max, 76
West, James, 76, 119–120
Wiener, Norbert, 185
Witchcraft, Navaho, 77
Women, role as corrective agents, 116; source of prestige, 101
Work, concept of, 113–114
Works Progress Administration, 65
World War II, service, 102. *See also* Veterans, loans to
Wylie, Phillip, 148

Zodiac, signs, 81, 86

Date Due